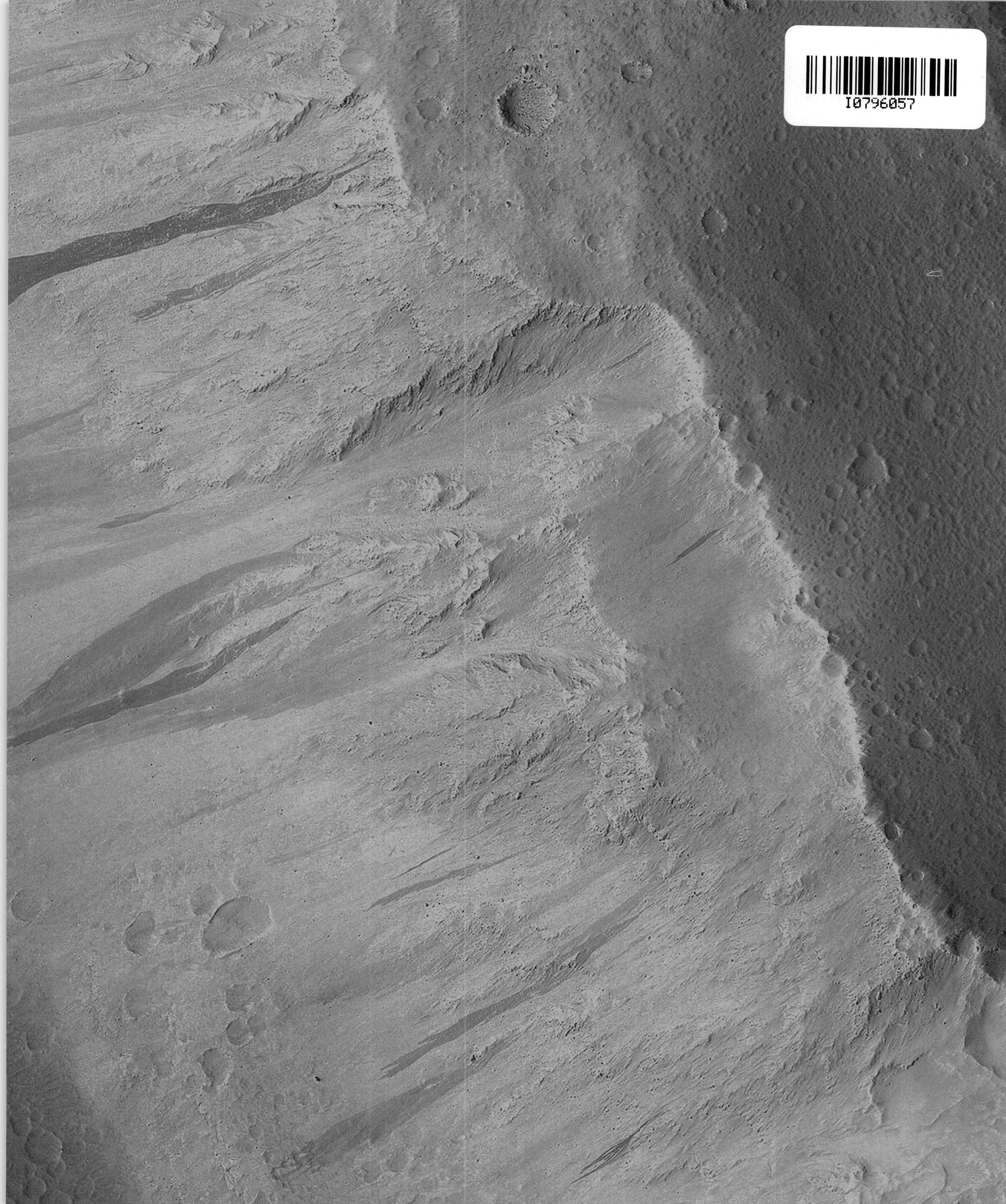

THE LAUNCH OF
ROCKET LAB

ELECTRON

Sir Peter Beck
Introduction

Peter Griffin
Text

THE LAUNCH OF

ROCKET LAB

Blackwell&Ruth.

"Somewhere between unlikely and impossible is where magic happens."

— Sir Peter Beck

Contents

Parts

Defying Gravity, Defining an Industry

As the clock struck midnight on November 25, 2024, Rocket Lab etched its name into the annals of space history.

In a feat never before accomplished, the company had just launched two rockets from different hemispheres within a twenty-four-hour period. The first Electron rocket, on a classified suborbital mission, thundered skyward from Wallops Island, Virginia. Its counterpart ignited the New Zealand sky from the Mahia Peninsula less than a day later, delivering five Internet of Things (IoT) satellites into orbit for its client Kinéis as part of the "Ice AIS Baby" mission.

This audacious double launch not only showcased Rocket Lab's operational prowess, but also symbolized the company's meteoric rise in the commercial space industry. Since Sir Peter Beck founded Rocket Lab in 2006, the company has achieved many other firsts.

It was the first private company in the Southern Hemisphere to reach space with the Ātea-1 launch in 2009. With the Rutherford engine, it created the first 3D-printed, electric pump-fed rocket engine. Rocket Lab established the first private orbital launch site with Launch Complex 1 (LC-1) in Mahia, New Zealand in 2016. Its first commercial rocket launch was from LC-1 in 2018. Rocket Lab was the first company to attempt to recover a rocket using a mid-air helicopter catch in 2022.

It has shared some history-making achievements with its customers, too, including the first fully autonomous pharmaceutical production experiment conducted outside the International Space Station with Varda Space Industries' "W-1" mission in 2023–2024. With NASA's "CAPSTONE" mission, Rocket Lab became the first company to build, launch, and operate its own small rocket and its own spacecraft to place a satellite into orbit around the Moon.

But building a successful space company—and a Nasdaq-listed one at that—isn't about being first, it's about being the best. In an industry where failure is measured in spectacular fireballs and success in silent streaks across the sky, Rocket Lab has consistently defied expectations.

Its tale is not just about rockets and satellites. It's a narrative of human ingenuity, perseverance, and the unyielding drive to push beyond the boundaries of what's possible.

As the undeniable leader of dedicated small launch with its Electron rocket, now the company is gearing up for its most ambitious challenge yet—the debut of Neutron, a much, much bigger reusable rocket to challenge the rockets owned and operated by American billionaires and offer a different way of sending satellites to space for new-space companies building the constellations of the future, and for interplanetary and national security missions for US federal agencies.

Neutron will also be the rocket that ferries Rocket Lab's own satellite constellation as the company aims for the ultimate endgame: building, owning, and operating its own satellite constellation to serves millions of people across the planet.

How did a self-taught engineer from New Zealand, a country without a space industry or any rocket launch capabilities whatsoever, become one of the world's most successful international space companies building and launching its own rockets and satellites to space?

It's a twenty-year-long story of self-belief, perseverance, and a commitment to excellence that's centered around Rocket Lab's visionary founder, but which includes a long and faithful cast of engineers, designers, investors, and business partners who have joined Sir Peter on the Rocket Lab journey.

This history of the first twenty years of Rocket Lab tells the stories of those people, from the engineers and Mission Managers with nerves of steel responsible for getting Electron safely off the launchpad, to the propulsion engineers who came up with the pioneering design for the Rutherford engine, to early investors like Vinod Khosla, who saw the potential in Sir Peter and his ambitious business plans.

The Rocket Lab story is far from over. Indeed, it feels like it's just beginning. Welcome to the chronicle of a company that didn't just reach for the stars—it built a highway to them.

Introduction

Sir Peter Beck

"Your son needs to have more realistic expectations."

Those were the words that greeted my parents on a chilly evening in the 1980s during a meeting with my high school career guidance counselor. Building rockets was apparently not on the pre-approved checklist of jobs they expected young men from Invercargill, New Zealand to pursue.

My parents smiled politely, shook the man's hand, and told me in the car on the way home to disregard anyone who told me to shy away from doing something just because it seemed impossible.

To be fair, the likelihood of a young kid from a small town in a non-space-faring nation, with no university degree, establishing a space company that built and launched rockets to orbit at a time when that very act was illegal under a myriad of international treaties was, admittedly, slim. But not impossible. Somewhere between unlikely and impossible is where magic happens. This is where Rocket Lab was born.

Like most companies that seem like overnight success stories, Rocket Lab began in a garage. Today, we're 2,500 people strong across three countries. We've launched missions to the Moon, built spacecraft for Mars, launched more missions to orbit than most space companies in history, and are building spacecraft constellations that will deliver critical services to billions of people, spanning communications, defense, environmental, and scientific capabilities. In terms of locations with the most active rocket launch sites, the order goes: America, China, then Mahia Peninsula on New Zealand's remote east coast, thanks to how often we fly our Electron rocket for our long list of mission partners.

As you flip through the pages of this book, you'll journey through the highs and lows, triumphs and trials that have marked Rocket Lab's path. You'll meet the incredible people whose passion and dedication have been the backbone of our success. You'll hear about the lessons learned along the way, the time I got escorted out of Edwards Air Force Base in California (I'm still not allowed back), and the near-death experiences the company faced many times in the early years.

This book captures the essence of Rocket Lab's spirit. It's not just a chronicle of our achievements but a testament to what can be accomplished when you dare to try and refuse to give up. It's about the countless hours of building, testing, and retesting, the sleepless nights before launches, and the jubilation of mission success. It's about the moments of doubt that are somehow overcome by unwavering belief and unshakeable teamwork.

Importantly, this book also examines the "why" of going to space. It explores the largely invisible but essential role it currently plays in your life, and sheds light on just how much of an impact it will come to have in the future—as well as Rocket Lab's role in ushering in that new era in the years to come.

To those reading who have been part of Rocket Lab's journey so far, thank you. To our team (and your ever-patient families), mission partners, customers, and investors—thank you for believing in our unreasonable vision.

I hope our stories spark curiosity and inspiration, and perhaps some new totally unrealistic expectations of your own, too.

"Your son needs to have more realistic expectations."

PART I

LOOKING UPWARDS

1

ELECTRON

EARLY

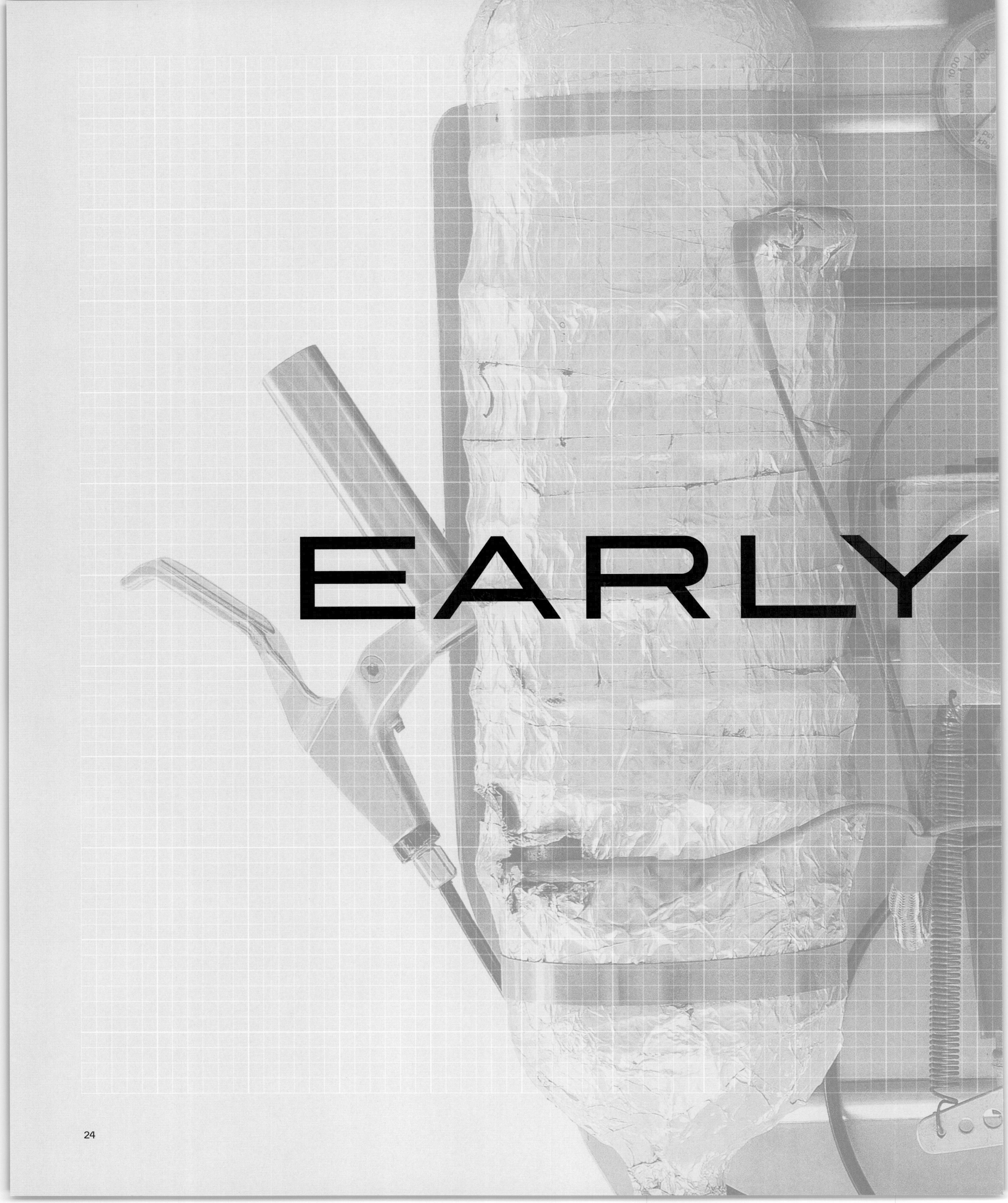

YEARS

Sir Peter Beck's fascination with space began in his childhood, under the starry skies of Invercargill, a city in the southernmost reaches of New Zealand where the rugged coastline meets the vast expanse of the Southern Ocean.

Often described as the "last bus stop" before Antarctica, which lies 2,612 miles (4,203 km) due south, the city of approximately 57,000 people is surrounded by lush farmland. The nearby Tiwai Point aluminum smelter creates some of the highest-quality aluminum in the world.

The Southland region was settled by predominantly Scottish colonists in the 1850s, a legacy reflected in city street names like Dee, Tay, and Spey. But the region was inhabited much earlier by indigenous Māori, who knew the area as Murihiku, meaning "the tail end of the land."

As a young boy, Peter's imagination was captivated by the universe, thanks to his father's influence. Russell Beck was not only an archaeologist, gemologist, museum curator, and artist but also a passionate amateur astronomer who influenced his son to look to the stars and consider his place in the universe.

Stargazing and Tinkering

Russell served as Director of Southland Museum and Art Gallery from 1976 to 1999. During that period, he introduced thousands of Southland children to astronomy with his homemade telescope, the lenses of which he had machined himself.

Peter started attending meetings of the Southland Astronomical Society at five years of age. "All I remember is sitting and listening to these woolly old men," he said. "I probably understood about 5 percent of it. But even the 5 percent I understood was just enthralling."

During his childhood, Peter would spend many nights in the observatory in front of the museum, helping defog the lenses so Southlanders could glimpse distant planets. Russell had artistic flair, creating numerous sculptures still dotted around New Zealand. His most notable creation is the giant cube outside the Southern Institute of Technology, designed to form an optical illusion and spur students to question everything.

Peter appreciated his father's artistic output, but wasn't compelled to become an artist himself, though this aesthetic heavily influenced the design of the Electron rockets Rocket Lab would produce decades later.

"I had no interest in the museum, I had very little interest in the art," he said. "I was interested in engineering, so if Dad was working late one night, I'd be deep in a gearbox somewhere."

It was during those stargazing sessions that Peter first began to think about rockets. Russell explained that each twinkling point of light could be a star with its own planets, and there might even be someone on a distant world looking back at them.

Russell had also dabbled briefly in rocket building in the mid-1950s, but pivoted to astronomy when, rather than heading skyward, his test rocket blew up.

"I had very little interest in the art, I was interested in engineering, so if Dad was working late one night, I'd be deep in a gearbox somewhere."

— Sir Peter Beck

Left: Nugget Point Lighthouse, Otago, New Zealand.

Right: A young Peter Beck rides an experimental bicycle.

“In the Beck household there really was only one rule and that was, ‘make an impact.’”

— Sir Peter Beck

Peter enjoyed school but was less interested in rote learning than in projects where he could apply knowledge. “If I could see how that knowledge was going to be applied, then I was all in there and loving it,” he said.

Space continued to fascinate him. When Halley’s Comet appeared in 1986, nine-year-old Peter became the resident expert in his class. Later, he learned much from his metalwork teachers at James Hargest College, where he was trusted with unsupervised access to the workshop.

Beck’s first major project was an aluminum bicycle. Peter’s mother Ann Beck, who was a schoolteacher, allowed Beck to experiment with the oven to heat the various components to allow them to fit, and then transfer the parts to the freezer to harden them.

“In the Beck household there really was only one rule and that was, ‘make an impact,’” he said. “So it was always an environment where we were encouraged to go big, and nothing was impossible.”

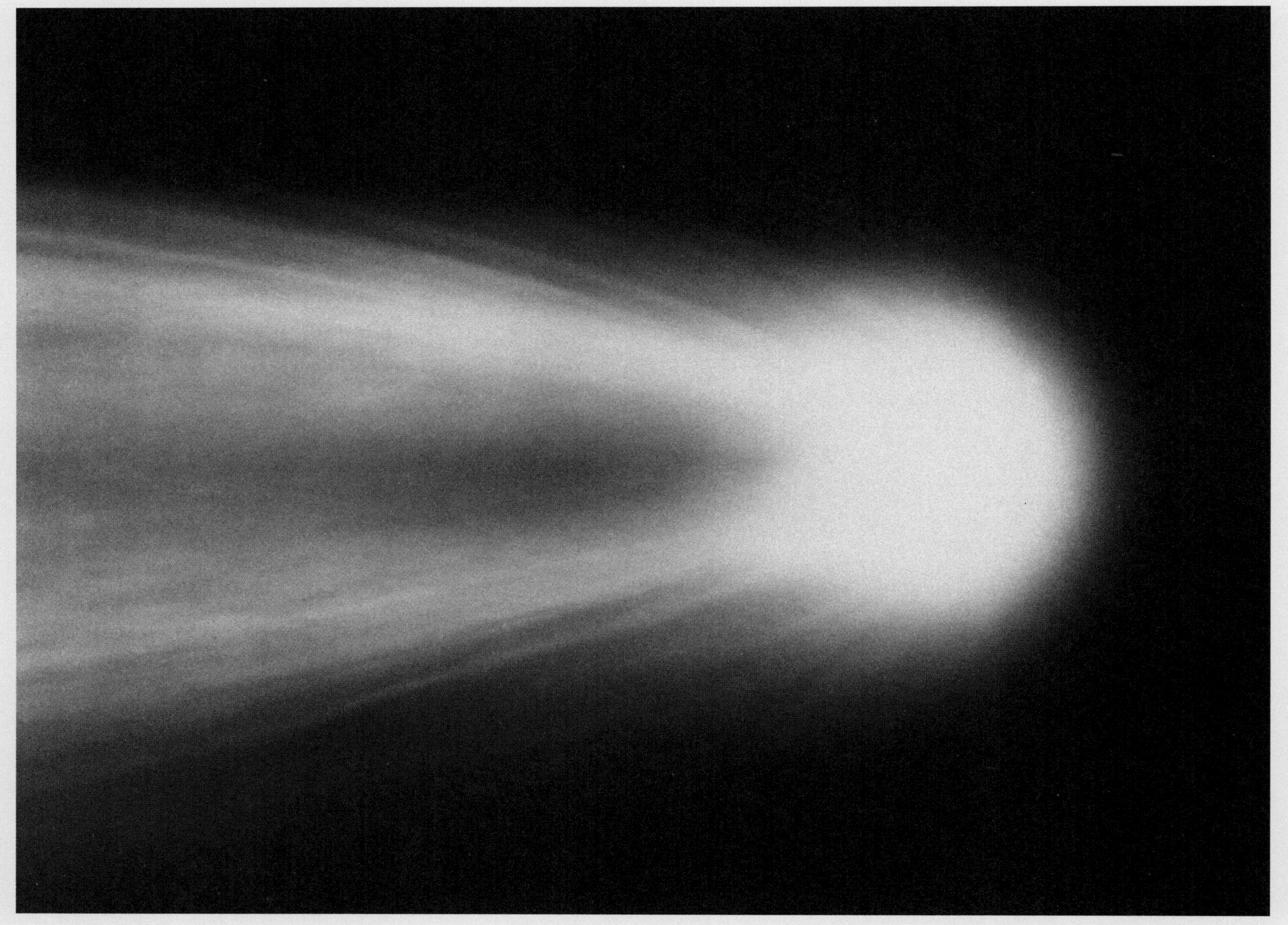

A Playground of Possibilities

Behind the Beck family home stood the workshop—a basic garage, but also a sanctuary where ideas could take physical form. Peter described an element of "brotherly tension" between the three boys due to their competitive natures, but emphasized they were "an extremely close family."

The workshop became his classroom, teaching him practical skills no traditional school could offer. With school friends, he built a motorized farm bike, beginning as a school project but pushing boundaries by sourcing a 200cc motor instead of the approved 100cc.

Peter moved on to his own car restoration, claiming a spot in the family workshop for a Mini he bought for NZ$300. "I drove it home and you could see the road through the floor," he said. Within six months he had almost entirely rebuilt it, adding a turbocharger for sports car acceleration.

As Peter prepared to leave high school, his passion for rocketry was no longer just a hobby—it had become a calling. He saw rockets as the ultimate engineering challenge, representing the pinnacle of human technological achievement.

He faced a pivotal decision—whether to pursue an engineering degree or seek immediate practical experience. He opted for the latter, applying for an apprenticeship in tool- and die-making rather than mechanics.

Dishwashers by Day, Rockets by Night

Peter Beck's journey to becoming a rocket engineer began with an unconventional approach to learning. As he told the magazine *NZBusiness*: "Ultimately, it doesn't matter if you go to university or not—you can end up in the same place. You just have a different approach to getting there."

In 1995, at seventeen, he joined whiteware maker Fisher & Paykel's apprenticeship program, moving to Dunedin to work in their Mosgiel factory. He felt like he'd "won the jackpot," gaining access to state-of-the-art equipment and knowledgeable engineers.

His talent stood out. Chris Staynes, the Mosgiel site's General Manager, described him as a "very, very bright" young engineer, "well above" the company's high standards.

While working on assigned projects, Peter utilized the workshop's resources for his rocket experiments. He often approached the engineering store for specialized materials: "I'd be trying to build a piece of my rocket and I'd go into the engineering store and say to Trevor, 'Hey Trev, can you get us a price for a piece of titanium, say yay by yay?' And he'd say 'Yeah, no problems, Pete,'" he recalled.

Left: Halley's Comet, 1986.

Above and right: Peter Beck's rebuilt Mini Cooper.

The Rocket Bike

One of Peter's most notable projects was a rocket-powered bike that could accelerate from zero to 87 mph (140 km/h) in under five seconds. During a demonstration at the Festival of Speed, he faced an unexpected challenge when the brake pads melted. He improvised, using wind resistance to slow down before applying the brakes.

A photo in the *Otago Daily Times* captured Peter in a red-leather bike suit, black helmet, on his yellow bike trailing a cloud of steam.

Below: Peter Beck's rocket-powered bike, capable of 0–86 mph (0–140 km/h) in under five seconds.

Left: Peter Beck goes racing.

Below: An experiment on rollerblades.

Overleaf left: An early Peter Beck invention: the Rocket Scooter.

Overleaf right and pages 24–25: The Rocket Backpack worn by Peter Beck during testing on rollerblades.

"Once you build an engine, how do you really verify how good it is? . . . strap it to your back."

— Sir Peter Beck

Things Snowballed a Bit
As Peter's ambitions grew, he began experimenting with hydrogen peroxide as a rocket propellant. Initially conducting tests in his house, he soon realized the dangers when he woke up with a severe headache from reactor fumes.

He explained his motivation: "Once you build an engine, how do you really verify how good it is? Being a little bit of a thrill-seeker and enjoying speed, the natural progression was to ride it, and then strap it to your back. These things sort of snowballed a little bit."

On the Same Wavelength
At Fisher & Paykel, Peter met his future wife, Kerryn Morris, a fellow designer. Their partnership would support him through future challenges. He explained: "My wife is infinitely practical like me and we're on the same wavelength for everything. So it makes it a very joyous and easy marriage when two engineers engineer together. Whether you're engineering a child or engineering life, it's just engineering."

Kerryn often acted as Chief Safety Officer for Peter's rocket bike tests. During one test, he experienced a direct short between the igniter system and the vehicle: "I had a direct short between this high-energy igniter system and the actual vehicle itself. So when I went to go and press the button, I completed the circuit, and it just belted me back onto the ground."

When he came to, Kerryn was standing over him laughing, thinking it was "the funniest thing she'd ever seen."

Peter completed his four-year apprenticeship in three years, quickly moving from the workshop floor to the design room. When Kerryn was offered a job as an oil field engineer in New Plymouth, Peter was ready for a change, and followed her to the new opportunity.

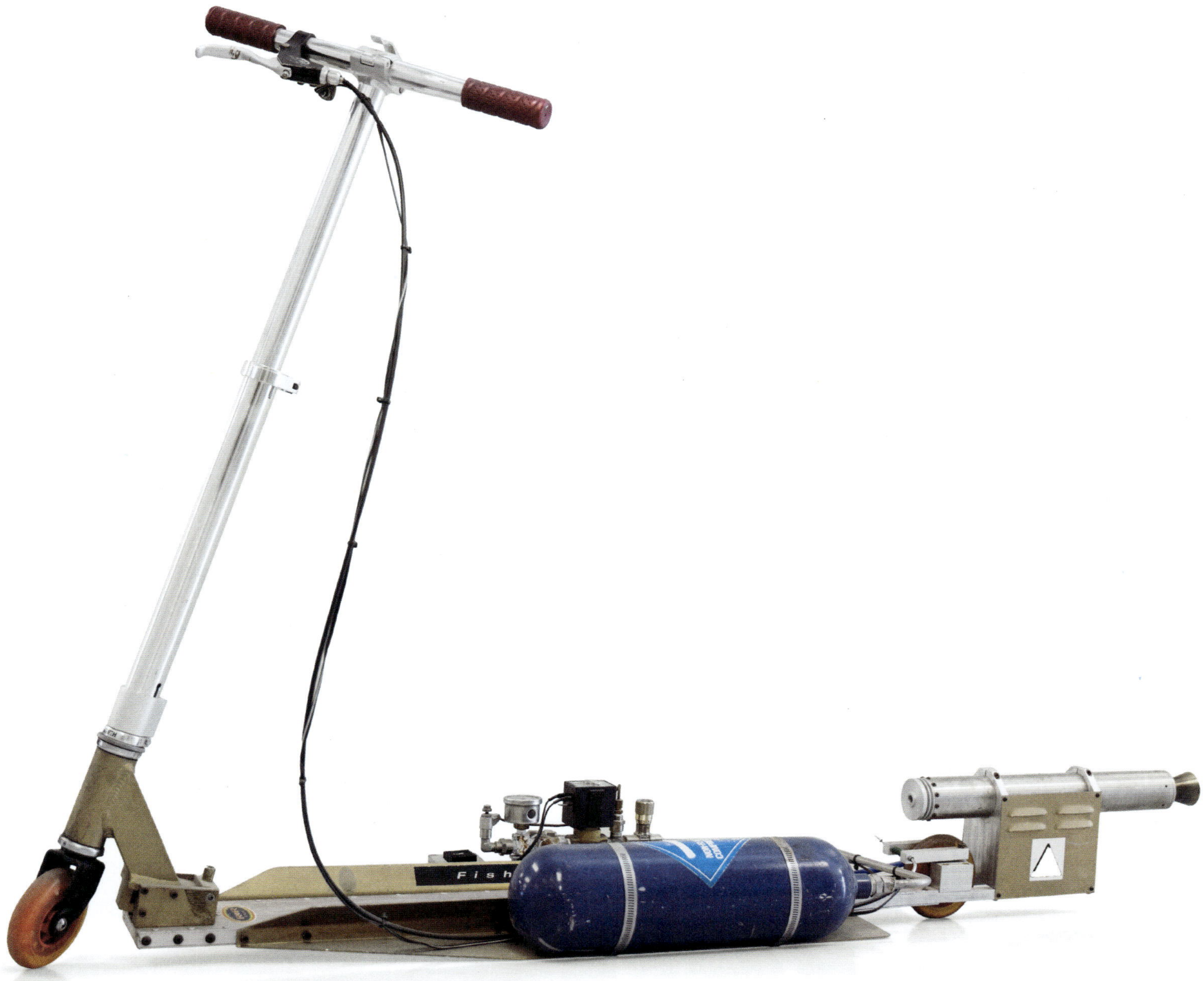
Fish

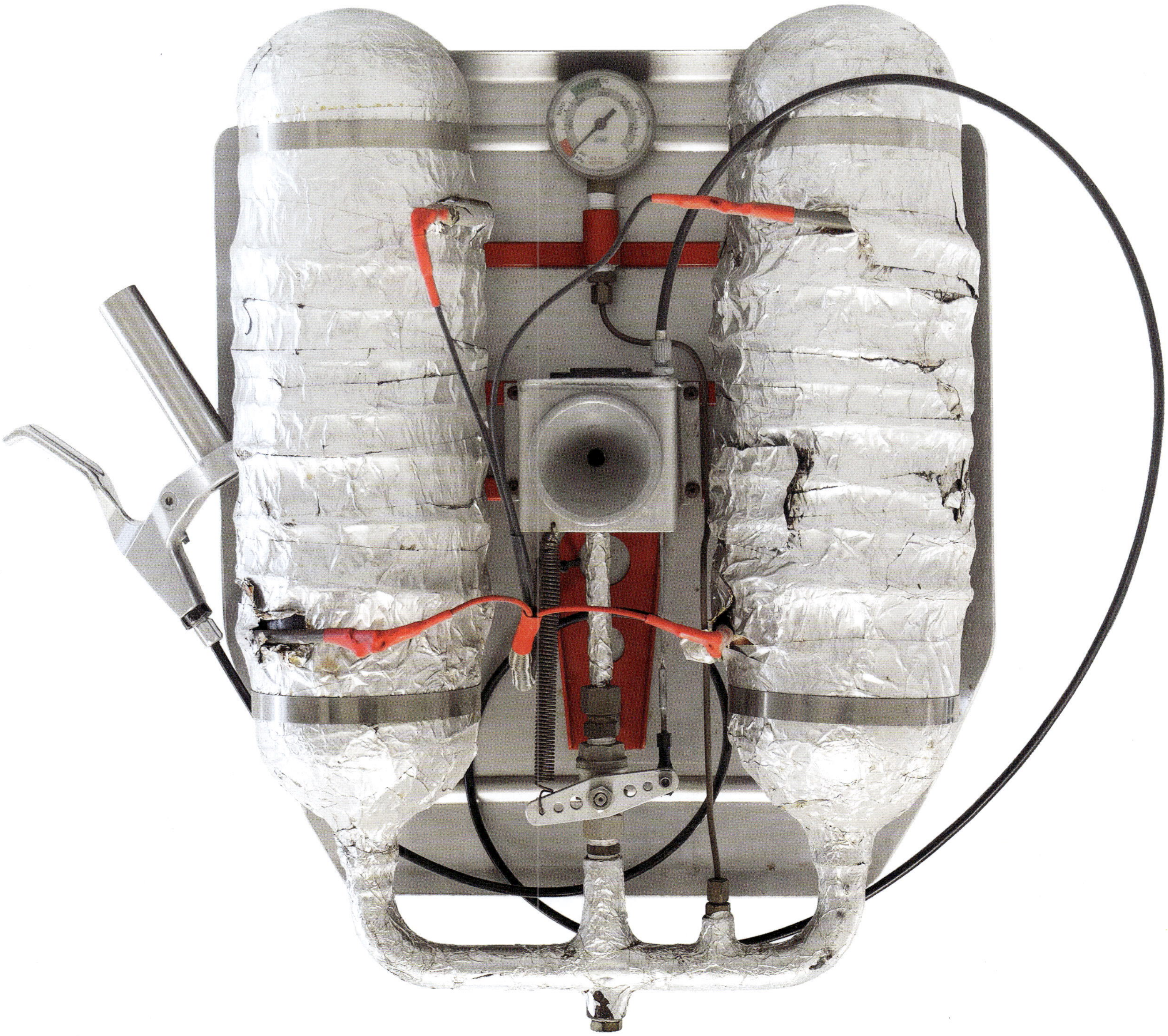

From Superyachts to Government Research

In 2001, after six transformative years at Fisher & Paykel, Peter Beck found himself working as a project manager for superyacht builder Fitzroy Yachts in New Plymouth.

"I don't think I enjoyed it that much," he said. "But I look back on that now super fondly, because it really did shape a lot of things. Having to keep everybody happy and drive a big project like that forward taught me a lot about people and people management."

At the time, Industrial Research Limited (IRL) was a state-run scientific institution with a laboratory in Auckland, New Zealand's largest city. Peter visited the lab in 2002 with a problem: developing a sound suppression system to cancel out superyacht engine noise.

In 2001, he landed a job at IRL (now Callaghan Innovation) and moved to Auckland where he could continue working on his passion—rockets. On his first day, he met Mark Battley, a Research Engineer and Carbon Composites Expert.

"He had this giant tank full of water with a little window on the side and a massive hydraulic ram on the top," Peter recalled. "Mark's like, 'Watch this' and he just slammed this composite panel into this water."

Mark was stress-testing carbon fiber materials for America's Cup racing yachts. Peter immediately saw potential for rocket

These pages: Peter Beck's pilgrimage to the United States.

construction and later had opportunities to examine America's Cup boats, learning about composite materials. His work varied widely—from testing wind turbine blades to solving problems for a ski racing boat.

"There was no solution, other than me hanging off the back of the boat with a laptop while this guy was just giving this boat absolute death to try and strain the gearbox to its maximum," he said.

Peter was perfect for IRL—he understood mechanics, had design skills, and could build things. He worked with software, superconductors, and high-performance materials. During evenings and weekends, he used the lab to develop rocket components and control systems.

Life was good. His wife Kerryn worked in the oil industry while Peter benefitted from the resources to build his knowledge and prototype rocket parts. He aimed to work for NASA eventually. Kerryn had given him a NASA Space Shuttle model that sat on his computer monitor.

"It was in my eyeline, that was my target," he said. The model still sits on his desk at Rocket Lab, where NASA is now a major customer.

A Moment of Clarity at 30,000 feet

In 2006, Peter joined Kerryn on a US business trip, planning visits to NASA's launchpad at Lockheed Martin's Cape Canaveral, Florida, location, and the Jet Propulsion Laboratory (JPL) in California. He'd been sending his resumé to these organizations.

As the tour progressed, he became depressed, realizing his chances at these institutions were minimal.

"It became super clear that a foreign national working in the space sector, which is a dual-use environment, was not a goer," he said.

Without US citizenship or a green card, he couldn't access NASA's technology. "Let alone a foreign national with no university degree from a country that has no space heritage whatsoever."

His JPL visit was particularly disappointing. "The average age at that point was fifty-two," he said. "Most programs were dying. All the cool stuff looked like it had been done. So I walked out of there thinking, 'This is all wrong, this is not a place that's going to suit me.'"

Ironically, fellow New Zealander William Pickering had run JPL during its 1960s heyday. Now it seemed bureaucratic and struggling.

The trip saw Peter attempt to visit Edwards Air Force Base in California's Mojave Desert, hallowed ground for aerospace nerds. Pulling down a long dusty road, Peter stopped his hire car just short of a guard house and hopped out with his camera to take some photos of the Edwards Air Force Base sign.

"Suddenly a guard appeared and grabbed my arm as I was taking the photo," said Peter. "He started asking me a lot of very awkward questions. Of course, the only identification I had on me was a New Zealand passport. So instantly, I'm a foreign national taking photographs at a military installation."

It was only by latching onto the guard's love for the *Lord of the Rings* movies that Peter was able to talk his way out of the situation. "He suggested I leave quickly and not come back."

But Peter would be back. Five years later he was invited to Edwards Air Force Base as a special guest. He proudly offered up his New Zealand passport as identification at the same guard house.

Visiting rocket startups in the desert, Peter realized they weren't much further ahead than he was.

"It became super clear that a foreign national working in the space sector, which is a dual-use environment, was not a goer, let alone a foreign national with no university degree from a country that has no space heritage whatsoever."

— Sir Peter Beck

"I started thinking about it logically, as engineers do, and there was only one obvious thing left to do. The logical thing to do was to start a company."

— Sir Peter Beck

"I can remember one startup and they were very proud to show me a torch ignitor that the US government had funded them to the tune of one million dollars to develop," he said. "They proceeded to show me this janky machine with rough edges screwed to a piece of plywood. I'm like, 'Man, if I'm the US government and someone presents that and I've spent a million dollars on it, I'd throw it back in their face.'"

Though inspired by meeting Ky "Rocketman" Michaelson, Peter knew a NASA career wasn't possible.

"Shit, what am I going to do?" he thought as he left LAX. "I started thinking about it logically, as engineers do, and there was only one obvious thing left to do. The logical thing to do was to start a company."

Halfway through the flight, he began mapping plans for his own rocket company on paper. He kept the name simple: Rocket Lab.

These pages: Peter Beck at the Kennedy Space Center in Cape Canaveral, Florida, US.

Overleaf: Peter Beck holds a model rocket engine.

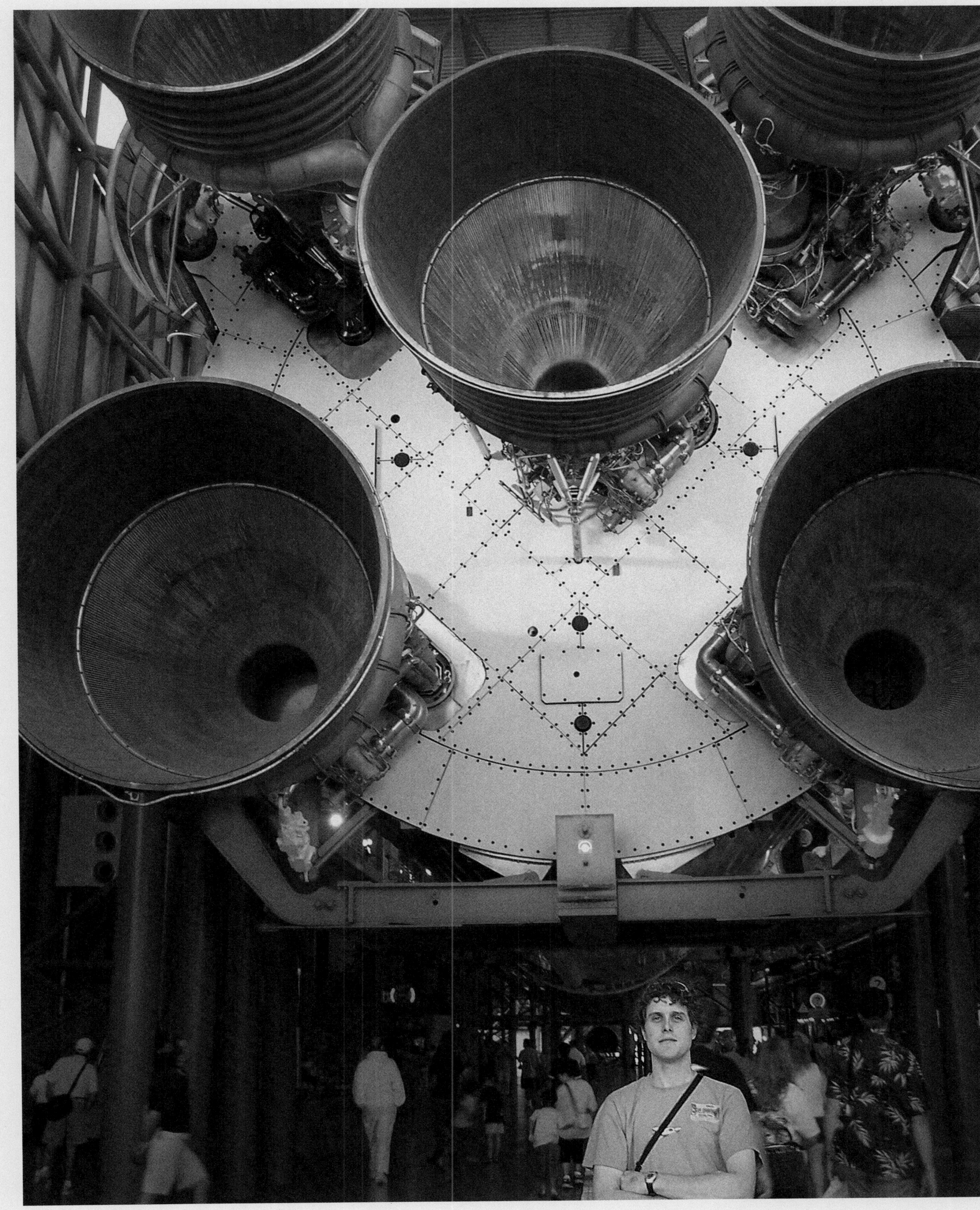

Rocket Lab is Born

Back in Auckland, Peter Beck handed in his notice at IRL. But he didn't move far, shifting down the hallway in the warehouse facility to his own private offices and renting access to the laboratory. He also didn't have to look far for an engineering collaborator.

One day outside the IRL lab, he approached Shaun O'Donnell, an R&D engineer he'd got to know well working alongside on various projects. He asked him if he wanted to join Rocket Lab. Shaun was an electrical engineer by training and not obsessed with space as Peter was. "I was just always interested in technology and doing new and different things," he said.

Having assisted numerous startups at IRL with their engineering needs, he wasn't intimidated by the prospect of leaving the security of the company.

"I kind of appreciated that model and Peter was super motivated. He worked fast. He was just a good engineer. When the opportunity came up, it was hard to resist."

Peter and Shaun got to work on designs for a small rocket, with Shaun taking responsibility for avionics—the electronic and software systems that would guide and control the rocket.

Peter now faced an urgent problem—Rocket Lab subsisted on his dwindling savings. New Zealand had no shortage of space-obsessed amateurs dreaming of launching rockets. But entrepreneurs with cash to back their dreams were thin on the ground.

In Search of Something Bigger

One day, he heard a successful software startup founder talking on the radio. Mark Rocket—who had, a few years previously, changed his surname by deed poll from Stevens—was now flush with cash.

He contacted Mark and the two hit it off. Mark put pre-seed funding into Rocket Lab in 2007, taking a 50 percent stake, and joining as a director.

Between 2007 and 2009, Peter used the small amount of cash Mark had invested to develop Rocket Lab's first suborbital rocket, Ātea-1. Ātea is the Māori word for space.

Shaun remembers the focus he and Peter brought to the challenge: "We picked the things that were the hardest first, and tried to break down the research into them initially. Because we had that small amount of money, it was about getting the most value out of it."

By early 2009, Rocket Lab had developed the key components for Ātea-1. Peter now turned his attention to a potential launch site.

He consulted a map of military operational zones around New Zealand. Great Mercury Island, off the east coast of the Coromandel Peninsula, caught his attention. He discovered Sir Michael Fay, a prominent New Zealand businessman, owned it.

"He would rent his house on Great Mercury out to billionaires and celebrities like Tom Cruise, so I rang the company that handled bookings. I told them, 'I don't want to book your house, I want to book the island to launch a rocket.'"

Michael was amused and agreed to meet him at the lab in Auckland.

"Michael was just there for fun; he thought it was going to be a good old laugh," Peter said. "But when he saw the rocket, what we were doing, and how serious we were, he was like, 'Yeah, let's do this. This is great!'"

Michael began preparing the way for his island to be temporarily taken over as a launch facility. "Michael gave us free rein. If we needed a helicopter, if we needed a boat, 'Here, just use it!'"

> "In the tradition of great New Zealand explorers, New Zealand, we are go for space."
>
> — Sir Peter Beck

Launch Day on Great Mercury

November 30, 2009 dawned with the slender, 19-foot (6 m) long, 132-pound (60 kg) Ātea-1 on its makeshift grassy launchpad. The island was a buzz of activity as Rocket Lab staff, journalists, and observers gathered.

"Sir Michael Fay really put on a lot of things for us," said Shaun. "I remember there were a lot of journalists there and a big party. I certainly hadn't experienced anything quite like that."

Peter went into that day quietly confident in what he and his tiny team had created. "Maybe there's a certain level of arrogance there," he reflected. "But you have to be confident in your workmanship, otherwise you would never do it."

An initial attempt to launch was scrubbed when the super-cooled fuel froze, leaving the rocket stuck to the launchpad. Peter hadn't planned for that eventuality. "We had to do a little bit of a dodgy thing, where we opened the main oxidizer valve without any of the ignition system within the engine activated," he said.

It worked, but he now needed a replacement air coupler valve. Michael dispatched his helicopter pilot to fly Peter to Whitianga to pick one up.

"This guy met us at the airfield with a check valve," Peter said. "I grabbed it out of his hand, jumped back in the helicopter, and buggered off. I completely forgot to pay him, which we had to sort out later."

The NZ$6 part was installed and Ātea-1 was refueled. Around 2.30 p.m. the last checks were made. In the final seconds before the countdown, Peter announced: "In the tradition of great New Zealand explorers, New Zealand, we are go for space."

He voiced the countdown himself, flicking control switches to trigger the pyrotechnic charge that ignited the rocket fuel. A sharp hiss signaled the rocket's lunge skyward.

"You fucking beauty! Yes!" Peter shouted as he watched the departing rocket.

"A lot of people were crying. It was really dramatic," Mark Rocket later told the New Zealand Press Association. "The power of the rocket is just incredible."

Ātea-1 burned for twenty seconds and reached a speed of Mach 5, or 3,106 mph (5,000 km/h). It ascended more than 62 miles (100 km), the internationally recognized altitude where space begins, called the Kármán Line, before flaming out, descending back over the ocean, and splashing down. A GPS tracker allowed the team to recover the rocket.

"There are heaps of times in Rocket Lab's history where all the chips are on the table," said Peter. "That was definitely one of them."

The local media coverage was extensive. But more important was word of the launch filtering back through Peter's network of aerospace contacts in the US. He had proven Rocket Lab's ability to achieve a lot in a relatively short period.

"In this industry, people will do a lot of talking, a lot of PowerPoints, and all the rest of it. Whereas I prefer to talk less and spend all that energy and time just making stuff and improving," he said.

"There are heaps of times in Rocket Lab's history where all the chips are on the table . . . That was definitely one of them."

— Sir Peter Beck

Above: The successful launch of Ātea-1 on Great Mercury Island off the east coast of New Zealand.

SUBORBITAL
SOUNDING
ROCKET
↓

ĀTEA-1

Name:	Ātea-1
Date:	November 30, 2009
Time:	04:20 UTC (16:20 NZT)
Site:	Great Mercury Island / Ahuahu New Zealand
	36.6274° S, 175.8025° E

Rocket Lab's Ātea-1 mission, launched on November 30, 2009, marked New Zealand's first successful space launch. Ātea-1, meaning "space" in Māori, was a suborbital sounding rocket designed to reach an altitude of 75 miles (120 km). The mission aimed to demonstrate Rocket Lab's capabilities in developing and launching cost-effective rockets for scientific research and commercial purposes. Ātea-1 carried a small payload to gather atmospheric data and validate the rocket's performance. This mission laid the groundwork for Rocket Lab's future technology, including the development of its Electron rocket.

KEY ATTRIBUTES	
Fuel	A hybrid fuel of polymer and liquid nitrous oxide
Thrust	1,550 lbf (pound force) for up to 14.5 seconds
Length	19.6 feet (6 m)
Diameter	0.49 feet (150 mm)
Mass	132 pounds (60 kg)
Payload	4.4 pounds (2 kg)
Altitude	90 miles (145 km)
Launch G-force	15 g's
Maximum velocity	Greater than Mach 5
Recovery	Land-or sea-based

America Comes Calling

After Ātea-1's success, Peter Beck needed to secure Rocket Lab's future. The rocket opened doors at the US Defense Advanced Research Projects Agency (DARPA), responsible for maintaining America's defense technology superiority.

"That really got things cooking at that point. All of the big defense contractors wanted a piece of that technology, and we became very relevant very quickly."

— Sir Peter Beck

Below: Viscous Liquid Monopropellant (VLM), developed by Rocket Lab.

Right: Peter Beck with the VLM test launch vehicle.

Opposite: Successful lift-off of the VLM test launch vehicle.

Peter met with David Neyland, Director of DARPA's Tactical Technology Office. Despite having his laptop and USB drives confiscated for security reasons, he made his pitch.

"I sat in Dave's office, which had no windows, with nothing, just me and Dave talking for twenty minutes of what was supposed to be a one-hour meeting," he remembered.

Beyond promoting Rocket Lab's sounding rockets, he pitched a new rocket propulsion concept—Viscous Liquid Monopropellant (VLM). This innovation combined solid and liquid propellant properties for a more efficient, safer fuel. VLM acts as a pseudo-solid until force is applied, then flows like a liquid, offering higher density, safer handling, and better combustion than traditional solid propellants.

David was interested, but Peter's foreign status presented challenges. However, a solution was very quickly arrived at, demonstrating DARPA's interest in VLM's potential for suborbital rockets and missiles. Rocket Lab received an R&D grant which was small for DARPA, but crucial for the cash-strapped startup.

The VLM work culminated in 2012 with a demonstrator rocket test on Great Mercury Island.

"That was a really important tech demo, because it was a very disruptive kind of thing, and we wanted to demonstrate it in the most disruptive way possible," Peter said.

With the US Department of Defense seeking AIM-9 Sidewinder air-to-air missiles, he assembled an invite list: "We invited DARPA and Lockheed Martin down, all of the people who were bidding on this program."

The test was successful, demonstrating throttling capabilities and 13 percent better engine performance than industry standards. US officials were amazed that a small New Zealand company achieved such improvements on a minimal budget.

"That really got things cooking at that point," said Peter. "All of the big defense contractors wanted a piece of that technology, and we became very relevant very quickly."

Rocket Lab patented their VLM system in 2012. Though not ultimately used by the military, the R&D contracts sustained the company and built credibility for future government contracts.

In December 2010, Rocket Lab won a contract from the Operationally Responsive Space Office (ORS), a Department of Defense unit established in 2007 to rapidly deploy small satellites for national security.

"They effectively funded us to do the initial conceptual design of Electron," said Peter.

Finally, his goal of developing an affordable orbital launch vehicle aligned with a funded project. Rocket Lab aimed to make space accessible through volume manufacturing of affordable launch vehicles.

Peter recognized the problems with traditional launches. They were expensive and infrequent, often with years between opportunities. He saw potential in serving the growing small satellite market with more affordable options.

Bus-Sized Behemoths to Palm-Sized Powerhouses

Gone are the days when satellites loomed like orbital buses as the only option for space-based communications and remote sensing, with eye-watering price tags in the hundreds of millions. Much like how phones have shrunk in size as technology has improved, today's CubeSats (small satellites used for space exploration and research) can nestle in the palm of your hand yet pack a technological punch. They have democratized space for everyone from government agencies to scrappy startups and curious university students.

While satellites shrank spectacularly, rockets remained stubbornly supersized, excruciatingly slow to build, and financially challenging. Small satellite creators faced a brutal choice: shell out a staggering $60 million for a dedicated launch on one of the world's few commercial rockets, or join a rideshare mission—enduring potential delays, limited launch windows, and often no precise control over their satellite's final orbital home.

Peter's Electron helped change this paradigm. Standing a mere 59 feet (18 m) tall—dwarfed by traditional launchers—it was intended to be able to blast off every seventy-two hours at an affordable price of just $5 to $7 million per launch.

This represented a cost reduction compared to SpaceX's Falcon 9 or Northrop Grumman's Pegasus launch. The small satellite revolution finally had its perfect launch partner.

Bloomberg journalist Ashlee Vance had questioned in his 2023 book *When the Heavens Went on Sale*: "Did the world need a rocket that could fly every week? Would it be a profitable business? Would our entire psychology around flying things to space change if a regular, cheap ride to the heavens existed? People were willing to pay to find out," he concluded.

Ashlee described Peter as an "engineering savant" who recognized that rockets were no longer exclusively for national space programs. "It was still rocket science, and it was hard. But the technology no longer had to live in a mythical realm populated entirely by geniuses. Rockets had become approachable."

Other entrepreneurs reached similar conclusions, particularly Elon Musk with SpaceX. When Ātea-1 made its maiden flight, he had already abandoned the Falcon 1 small-lift vehicle after three failures in five attempts. In 2009, SpaceX shifted focus to the larger Falcon 9, creating a market gap for Rocket Lab.

"I had no idea that Elon even existed," Peter admitted. "In those times, SpaceX looked no different to one of those little shops out in the Mojave Desert."

Time to Run Free

While Peter envisioned Electron, Rocket Lab's funding came primarily from defense-related projects—a situation that troubled cofounder Mark Rocket. By 2010, with Peter negotiating military contracts, Mark decided to leave.

"I have tremendous respect for Mark, but I needed to run free," Peter explained. "I needed to bet the company that minute of that day on a project and make commitments and get well over my skis. I needed absolute freedom to run."

"The reality is that the space industry is a dual-use industry. Mark was uncomfortable doing any work with the military. But the military created Rocket Lab and nurtured it in the very early years of its formation," he added.

In 2011, Mark sold his 50 percent share to Peter, giving him full control, but creating financial strain. Besides small New Zealand government grants, Rocket Lab relied on US R&D contracts.

In 2013, the company moved its registration to the US, facilitating work with the country's defense clients and NASA, while enabling access to key US space technologies for Electron's development.

Peter then accelerated development of Electron's key components: lightweight carbon composite structures, Rutherford engines, and advanced avionics systems.

From Bootstrapping to Sand Hill Road

In the early days of Rocket Lab, Peter Beck faced a critical challenge that many ambitious startups encounter: securing the necessary capital to transform a visionary idea into reality.

By 2011, he had raised the best part of $2 million in grants and R&D contracts, mainly from US defense agencies and aerospace companies. He'd parted ways with seed investor Mark Rocket in an amicable split that left him once again with 100 percent of Rocket Lab. But he was rapidly running out of money.

On a visit to the US, Peter visited Jim McCoy, the successful Silicon Valley investor he'd first met in New Zealand when McCoy visited as a guest of the New Zealand government to mentor local entrepreneurs. The soft-spoken yet shrewd investor had become a valuable sounding board for Beck.

"I stopped off at Jim's house and was like, 'I've got to make something stick here, Jim,'" said Peter.

"Peter, you're too nice," Jim told him firmly. "You have to take off the nice guy Kiwi hat. You have to be a bit of a wanker," he advised.

At the time, Peter was holding discussions separately with several US rocket companies about jointly developing solid propellant systems. He urgently needed a deal that would net Rocket Lab some much-needed cash, so took the advice and injected some urgency into the negotiations.

"I was a bit like a used car salesman at that point, playing one company off against the other," he remembered. "In the end, I did a half-million-dollar deal with Aerojet Rocketdyne. They paid me to stop talking to everyone else. That was one of the most important deals, because that saved the company."

Aerojet Rocketdyne had a long association with NASA, including providing the Rocketdyne F-1 engines that powered the Saturn V rocket for the Apollo missions, and later, the RS-25 engines for the Space Shuttle program.

Peter banked the check, and engineers from Aerojet Rocketdyne traveled to Rocket Lab to witness engine tests. Again, he was bootstrapping the company from one contract to another.

This approach allowed Rocket Lab to maintain control and develop its core technologies, but it also limited the pace of growth and innovation. By 2013, Peter realized that to achieve his ambitious goals for Rocket Lab, namely the development of the Electron rocket, he needed to secure substantial funding.

He set his sights on the epicenter of tech investment: Silicon Valley.

Right: Vinod Khosla, Khosla Ventures.

A Three-Week Gamble

Peter gave himself an arbitrary three-week deadline in Silicon Valley to secure the seed funding he needed for the next phase of Rocket Lab's growth. He aimed high, targeting a handful of top venture capital (VC) firms.

"We thought, 'We want to back this guy. We can't assess the probability of success, but it's worth taking a bet on him and on this large vision.'"

— Vinod Khosla

"I'd figured out by this stage that if you can get tier one investors, meaning the top of the venture capital tree, then by proxy other people want to invest in you, and it makes your subsequent rounds easier," he said.

He ended up securing meetings with three VC firms. Among them was Khosla Ventures, the Menlo Park-based company founded by Vinod Khosla in 2004. Khosla was a towering figure in California's VC community. He cofounded Sun Microsystems in 1982, with $300,000 in seed funding from Kleiner Perkins Caufield & Byers, one of the Valley's most established VC firms.

After establishing Sun Microsystems as one of the most important technology companies of the 1980s and achieving over $1 billion in annual sales just five years after its launch, Khosla left to join Kleiner Perkins as a general partner. What followed was a string of successful investments by Khosla in everything from video games to semiconductors.

Khosla Ventures has invested in startups like Impossible Foods, Instacart, Affirm, DoorDash, Square, and OpenAI. It also had a New Zealand connection before Beck knocked on its door. Khosla was an early investor in LanzaTech, an Auckland-based company that developed a pioneering method of using microbes in a bioreaction chamber to convert carbon emissions from industrial plants into biofuels and chemicals.

LanzaTech's cofounder Dr. Sean Simpson was on the same startup journey as Peter, and offered to make an introduction to Khosla.

"I still have the email Sean sent me about Peter and Rocket Lab," Vinod remembered. "We had worked with Sean and had a lot of respect for him, so we took it seriously."

Khosla Ventures also had space sector opportunities on its radar. In 2009, it had invested $3 million in satellite imaging company Skybox Imaging, adding a further $18 million alongside Bessemer Venture Partners two years later. Skybox Imaging would go on to raise an additional $70 million in a Series C funding round, before being acquired by Google in 2014 for $500 million.

Now Peter, the innovative engineer with a dream of democratizing space access, faced Vinod, a seasoned investor with a track record of identifying and nurturing groundbreaking technologies. The stakes were high, with Peter seeking $5 million in funding.

The First Big Pitch

Both men clearly remember the pitch meeting at Khosla Ventures' Sand Hill Road offices, where Peter literally laid his cards on the table.

"There were probably half a dozen sheets of paper that he rolled out on our conference table," said Vinod.

Among them was a blueprint for the Electron rocket.

"I had an electric turbopump, that we had kind of roughly tested, in a suitcase, and a very, very rudimentary pitch deck, and I just said, 'I'm going to build a rocket,'" recalled Peter.

He had no official aerospace industry lineage to draw on, nor a team of engineers with degrees from MIT or Stanford to pad out their resumés. That proved to be an advantage in the boardroom of Khosla Ventures. Vinod describes himself as a techno-optimist who believes that only the "improbable" ideas are important.

"One of the things I've learned is that experienced people seldom innovate," he explained. "So I only focus on people who are inexperienced. If Peter had hired a bunch of NASA and JPL people, it would have cost him five times as much."

As Vinod and Sven Strohband, a partner in the firm with an engineering background, listened to Peter's pitch, they found themselves impressed at his clarity of vision and ability to answer the questions they peppered him with.

"We thought, 'We want to back this guy,'" said Vinod. "We can't assess the probability of success, but it's worth taking a bet on him and on this large vision."

Peter nailed the pitch, but now the firm's partners went straight into a meeting to decide whether to make him an offer. Sven emerged from that meeting soon after, and gave him the good news—Khosla Ventures was keen to proceed with a $5-million-dollar seed investment.

Peter was exhilarated, but on hearing the equity stake in Rocket Lab he'd need to give up in exchange for the money, something took hold of him. "Words were just coming out of my mouth. I couldn't stop. It was 'Oh, that's great Sven, but that's not really what I'm looking for, it needs to be *this*.'"

Sven was taken aback. The offer was good, and they both knew it. Nevertheless, he headed back to the waiting partners with Peter's request.

"The deal was right within where I expected it to land," Peter reflected over a decade later. "It took me years to figure out that I think I said that because no good CEO is worth their salt if they accept the first deal," he reasoned.

As the minutes passed, he sat sweating, trying to figure out how he could walk back his rejection of the offer. "Then Sven came into the meeting room and said, 'Yup, we'll do the deal,'" he said.

Sticking to his guns saw Peter secure the funding and avoid a greater level of dilution of his stake in Rocket Lab. A period of due diligence followed, during which contacts from NASA and defense companies Rocket Lab had worked with over the years gave glowing recommendations.

Bouncy Balls and Turbopumps

Khosla Ventures' initial investment proved to be just the beginning of Rocket Lab's venture capital-powered funding journey. Between 2015 and 2021, when Rocket Lab listed on the Nasdaq stock exchange in New York, the company raised four more rounds of VC funding worth more than $280 million.

The Series A funding round in 2014 saw Peter back in pitching mode again and employing more unconventional methods to appeal to prospective funders. This time it was David Cowan and colleagues from Bessemer Venture Partners, another giant of Silicon Valley VC. David had been a board member of Skybox Imaging and had a close relationship with Vinod Khosla, so was well briefed on Rocket Lab. Still, he was skeptical of its chances of success.

"He was pretty rude about it," Peter recalled of his first meeting with David. "He said, 'Tell me why on earth I should invest in you? This sounds nuts.'"

Peter unzipped his suitcase and removed the new version of the electric turbopump on the table, then set about explaining the edge the electronically fired Rutherford engine would give Rocket Lab. The pitch had the same impact, but Peter saved his masterstroke for last.

He had filled his luggage with a sack full of cheap plastic bouncy balls. He was seeking a way to convey the market opportunity that lay ahead of Rocket Lab. "At the end of the pitch, I lifted the sack above my head, pulled the strings and just dumped this entire sack of bouncy balls on the big, long boardroom table," he said.

The bouncy balls had more bounce than he expected. "They were just bouncing everywhere, landing in people's coffee. It was carnage," he remembered. "Once it calmed down, I picked up a bouncy ball and I said, 'Each one of these balls represents a satellite that needs to be launched in the next five years. And each one of these bouncy balls is worth somewhere between $5 and $7 million dollars.'"

Deal done. His original strategy of targeting three top-tier VC firms had paid off. Vinod's reputation and network in Silicon Valley opened doors to other investors and allowed Peter to tap specialist expertise to go with the investment dollars.

Along its funding journey, Rocket Lab also received investment from the likes of New Zealand early-stage fund K1W1, Greenspring Associates, DCVC (Data Collective), Promus Ventures, Lockheed Martin, Future Fund (an Australian sovereign wealth fund), and New Zealand's Accident Compensation Corporation (ACC).

The capital injection came at a critical juncture in Rocket Lab's journey. By 2014, the company had developed most of the technologies necessary for their first test flight, which was planned for late 2015. Khosla Ventures' investment provided the resources needed to move from the conceptual and early development stages to actual rocket assembly and launch preparations.

Sven and David would join the Rocket Lab board, and over a decade on from that first pitch meeting, Vinod still enjoys regular dinners with Peter, who he describes as "meticulous" when it comes to Rocket Lab's engineering, but with the requisite boldness of vision and ability to overturn decades of space industry orthodoxy.

"We romantically fall in love at our firm with the loners with big visions, trying to do the audacious with a small amount of money," he said. "Almost all the other small rocket companies have taken a different approach, in my view."

"At the end of the pitch, I lifted the sack above my head, pulled the strings and just dumped this entire sack of bouncy balls on the big, long boardroom table. They were just bouncing everywhere, landing in people's coffee. It was carnage."

— Sir Peter Beck

Peter Beck's Fund-raising Approach

Rocket Lab's experience with Khosla Ventures offers valuable lessons for other startups, particularly those in capital-intensive industries like aerospace:

1 **Vision and expertise matter:** Beck's clear vision for Rocket Lab and his technical expertise were crucial in convincing Khosla to invest.

2 **Timing is critical:** seeking funding when the company had made significant technical progress but needed capital to move to the next stage was a strategic decision.

3 **The right investor can open doors:** Khosla's reputation and network in Silicon Valley provided Rocket Lab with more than just capital.

4 **Be prepared to think globally:** Beck's willingness to look beyond New Zealand for funding was essential to Rocket Lab's growth.

PART II

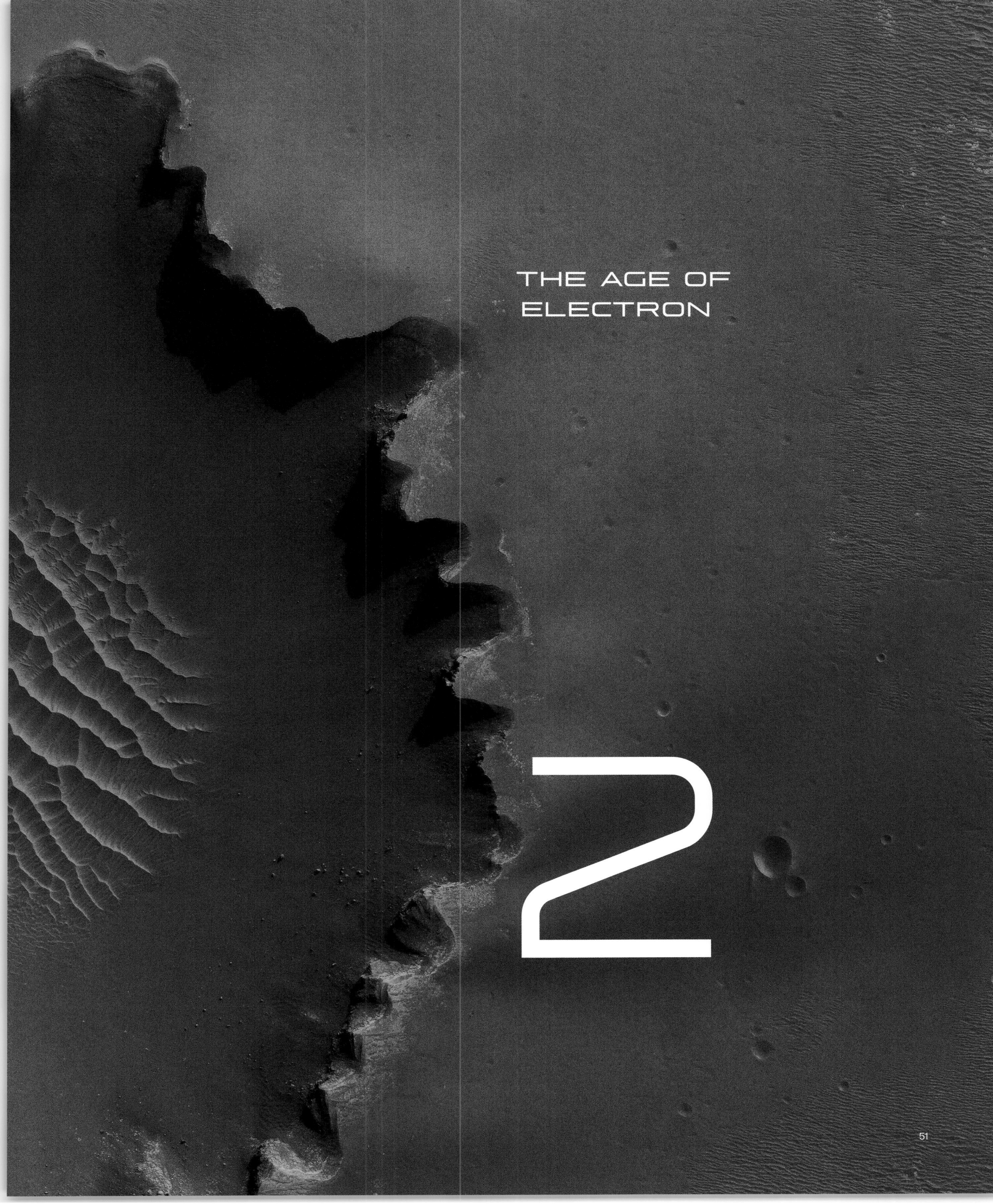

THE AGE OF ELECTRON

2

RUTHE

RFORD

Rutherford—The Heart of Electron

Every couple of years, classic planes take to the sky over Central Otago in New Zealand's South Island as part of Warbirds Over Wanaka, an airshow that has been delighting aviation enthusiasts since 1998.

"We haven't the money, so we've got to think."

— Ernest Rutherford

Lachlan Matchett, Rocket Lab's Deputy Chief Engineer, can't remember exactly the year he first made the three-hour trip south from his home in Palmerston to attend the show. What he does remember is the sound of it.

"Hearing those jet aircraft come past at high speeds, just all that noise and power really resonated with me. That got me into it," he said.

What followed was an infatuation with engines, then rockets, a trajectory that mirrored Peter Beck's, though while the latter had full use of a garage to tinker on cars and rocket-propelled bikes, Lachlan was relegated to the garden shed. The common denominator was parents willing to let their kids experiment and follow their curiosity.

"At school, I was pulling apart turbo chargers for cars and trucks and making them into jet engines," Lachlan recalled. "I found it was quite hard to make something really nice, because it needed a lot more precision manufacturing that I didn't have access to."

Rockets were different to jet engines. Noisier, simpler, requiring less precision. Following high school, Lachlan took a common path for those with an engineering bent, a Bachelor of Engineering degree at the University of Canterbury in Christchurch, one of the country's leading engineering schools. Aerospace, rockets, and gas turbines occupied his undergraduate studies, including designing and launching a research sounding rocket that reached a height of 6.2 miles (10 km).

With an offer to work at Rocket Lab as a contractor while he completed his Master's degree in trajectory optimization, Lachlan dropped everything and headed for Auckland. He was twenty-two, a fresh-faced university student in a new town, about to lead development of the engine that would propel the Electron rocket into space.

In all of the design iterations and prototyping Peter and his tiny team had been carrying out since the successful launch of Ātea-1 in 2009, much of them with the backing of US defense industry grants, was a new type of rocket engine they hoped would set Rocket Lab apart. At the time, most engines powering orbital launch vehicles used gas generator pumps. A portion of propellant would power a gas generator, which would drive turbopumps to feed the main combustion chamber where the propellants were ignited, creating thrust.

Peter instead wanted to use a simpler lithium-ion battery-powered electric turbopump, removing the need for a gas generator or staged combustion systems. It would be more efficient, simpler, and compact. Reducing the number of components would make the engine more reliable and easier to manufacture.

Electric pumps also allow for precise control over the propellant flow rates, enabling better throttling capabilities, and improved engine performance. This level of control is particularly valuable for the Electron rocket, which requires precise orbital insertions for its small satellite payloads. The batteries would add weight but allow for all of the propellant to be used for thrust generation.

"We saw this electrically driven system as a really good point of difference, something that could be brought to market faster," said Lachlan. "That's really what set up the idea for an electric rocket."

Inspired by a Nobel Winner

Peter already had a name in mind for the engine—Rutherford, in tribute to New Zealand's most famous scientist, Sir Ernest Rutherford. Born in Brightwater, a sunny town just outside Nelson at the top of the South Island of New Zealand, in 1871, the Nobel Prize winner undertook a series of groundbreaking experiments at the University of Manchester, culminating in him "splitting the atom" in 1917.

"All science is either physics or stamp collecting," Rutherford has been quoted as saying, a suitable utterance from the godfather of nuclear physics. But another saying Rutherford had about his small island homeland particularly resonated with Peter: "We haven't the money, so we've got to think."

The philosophy still applies to Peter's ethos at Rocket Lab. "We had no money, and even today, we are never going to outspend our two biggest competitors, Jeff Bezos and Elon Musk."

The Rutherford engine would embody that inventiveness and give Rocket Lab an edge in the commercial space race.

This page: A vacuum-optimized Stage 2 Rutherford engine donated to the Smithsonian National Air and Space Museum in 2022.

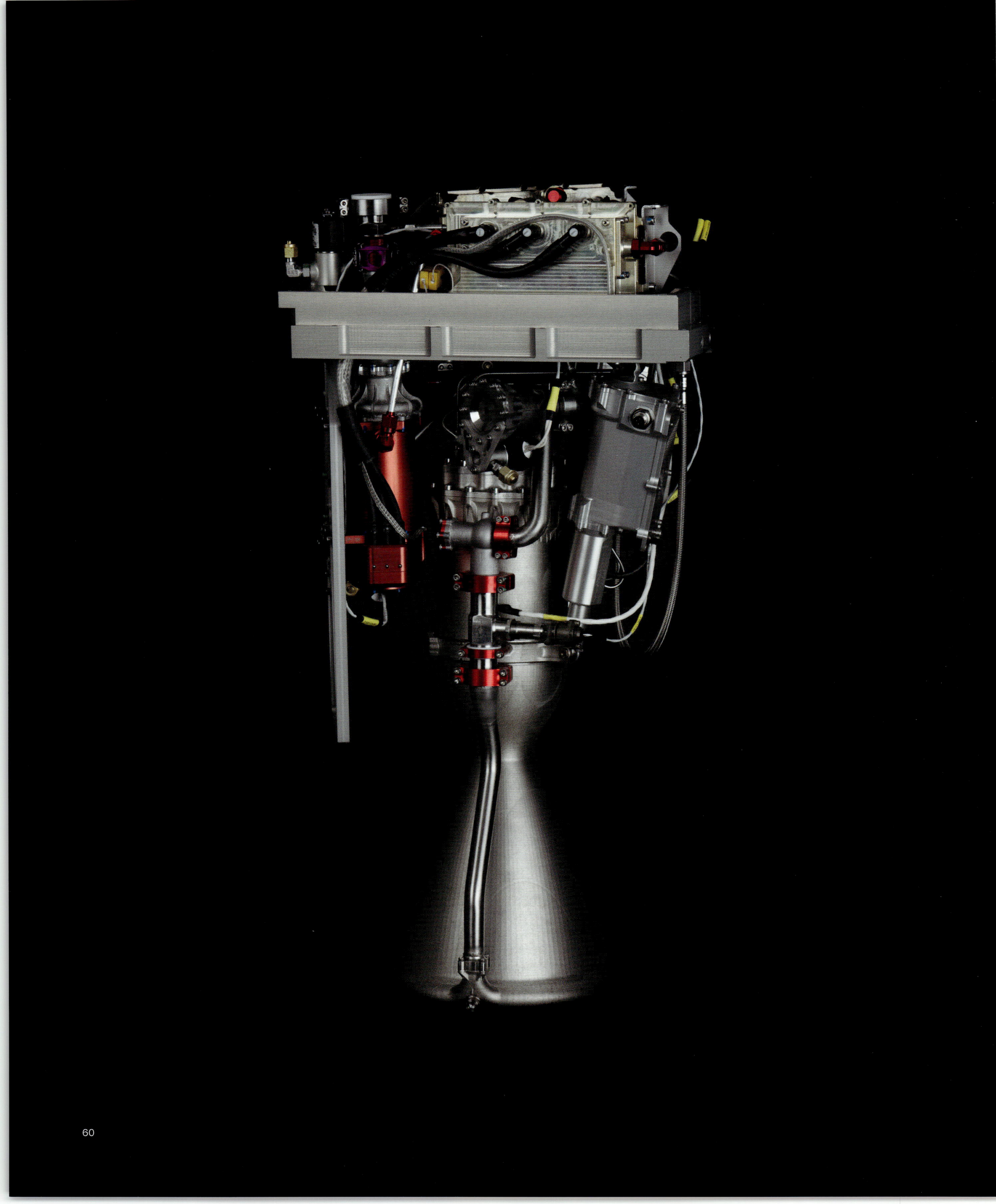

Within weeks of Lachlan's arrival at Rocket Lab, Peter was already preparing to leave for the US with the aim of securing funds from a Silicon Valley-based venture capital firm that would allow him to develop Rutherford and the Electron launch vehicle it would power. He needed something tangible to show the Valley's notoriously hard-nosed capitalists. Lachlan set to work building a rudimentary prototype.

"We used a drag racing fuel pump that we brought from the US. It looked all nice and red, anodized aluminum," he said. "We got an Ātea-1 injector that we pumped water through, and essentially hooked this whole chain up to show that we could make an electric pump-fed rocket engine work. That was the baseline for Rutherford."

By December 2013, the team was hot-firing engines and had the electric turbopumps running. But it took a further two years of development for Rutherford to deliver the performance and reliability Peter sought. The team was partially flying blind throughout, thanks to International Traffic in Arms Regulations (ITAR), a set of US Department of State regulations which governs the export of US military defense technology.

Under ITAR, design and engineering information relating to rockets, missiles, and space launch vehicles is carefully censored to avoid technology falling into the wrong hands. Employees from NASA, Boeing, or McDonnell Douglas weren't simply allowed to quit their job and move across the Pacific to New Zealand to work on rocket engines and launch vehicles for Rocket Lab, taking all of that sensitive knowledge with them.

"The toughest thing at the time was not being able to hire anyone from the US to help us, particularly with engines, but it was the same with the launch vehicles," Lachlan said.

There weren't even many textbooks the team could draw on; the engineers instead having to join the dots themselves and learn through trial and error. They puzzled over fluid flows and experimented with centrifugal pumps as they closed in on the ideal design.

> "At school, I was pulling apart turbo chargers for cars and trucks and making them into jet engines."
>
> — Lachlan Matchett

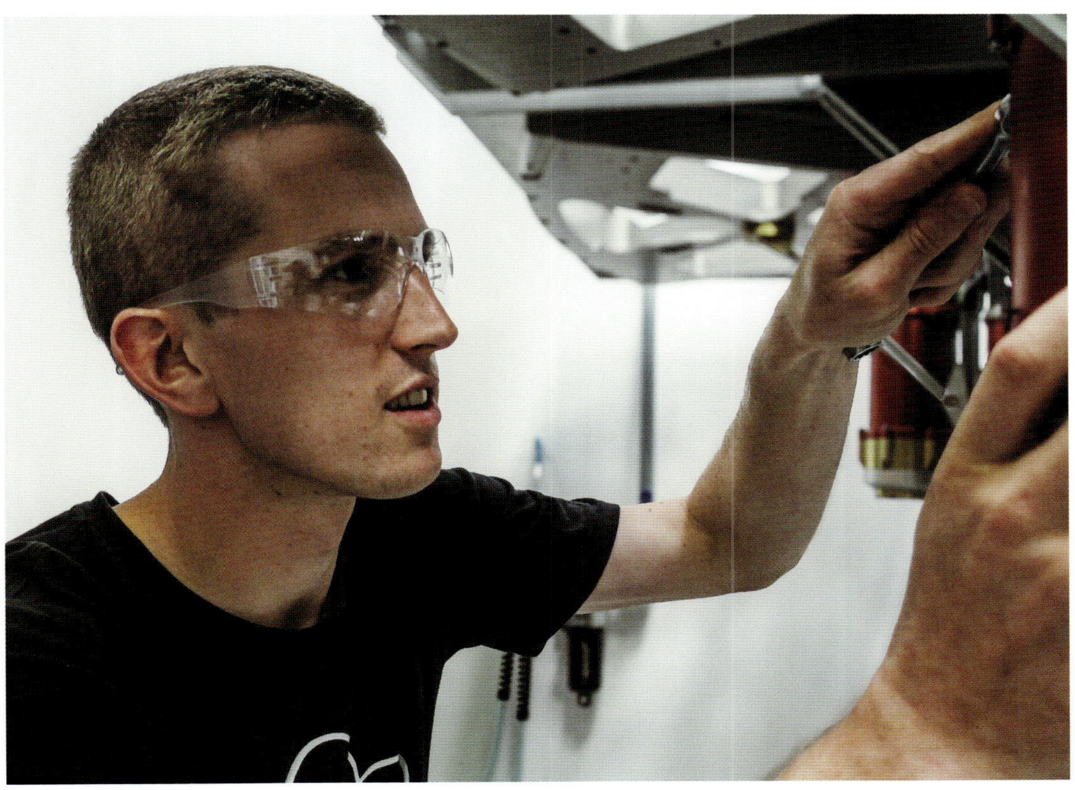

Opposite: A Stage 1 Rutherford engine.

Left: Lachlan Matchett, Deputy Chief Engineer, Rocket Lab.

Overleaf left: Rutherford engine testing, otherwise called a "hot fire".

Overleaf right: Engine testing of nine Rutherford engines simultaneously.

RUTHERFORD

The Rutherford engine is an innovative liquid-propellant rocket engine that uses electric pumps to feed its propellants, a first in the aerospace industry. The engine's electric pump-fed design reduces complexity and increases performance, enabling rapid manufacturing and cost-effective launches. Named after New Zealand-born physicist Ernest Rutherford, the engine is 3D-printed, enhancing production speed and precision. Rutherford engines power both the first and second stages of the Electron rocket and play a crucial role in Rocket Lab's mission success.

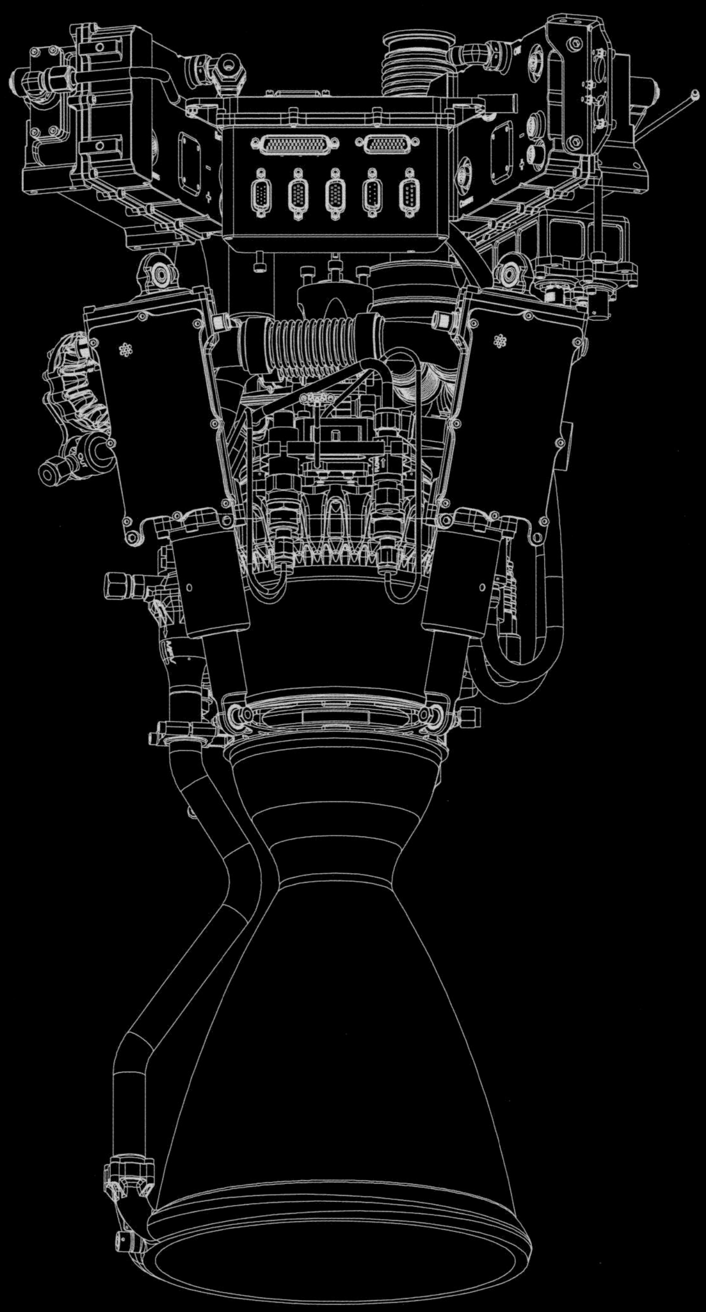

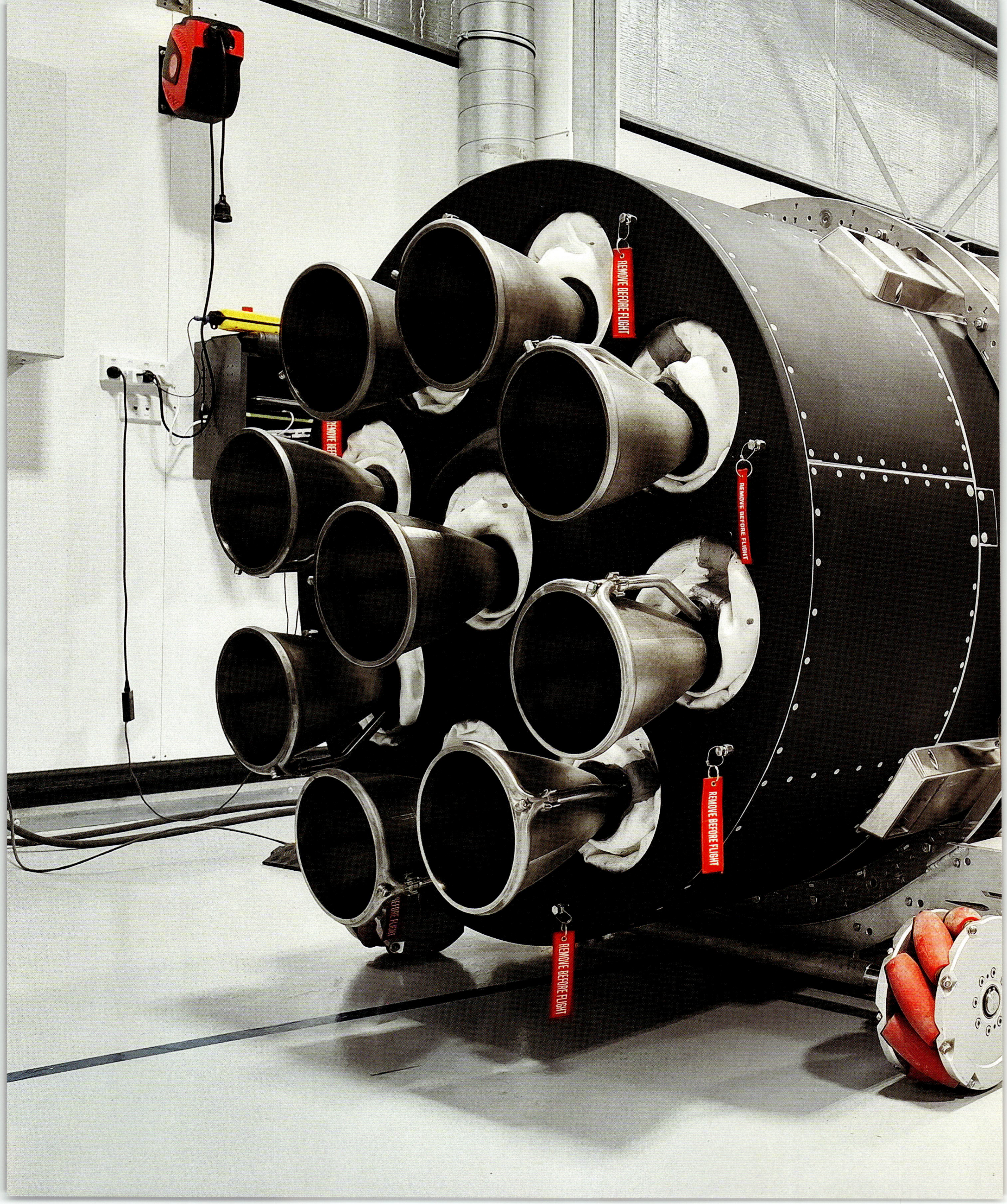
REMOVE BEFORE FLIGHT
REMOVE BEFORE FLIGHT
REMOVE BEFORE FLIGHT
REMOVE BEFORE FLIGHT

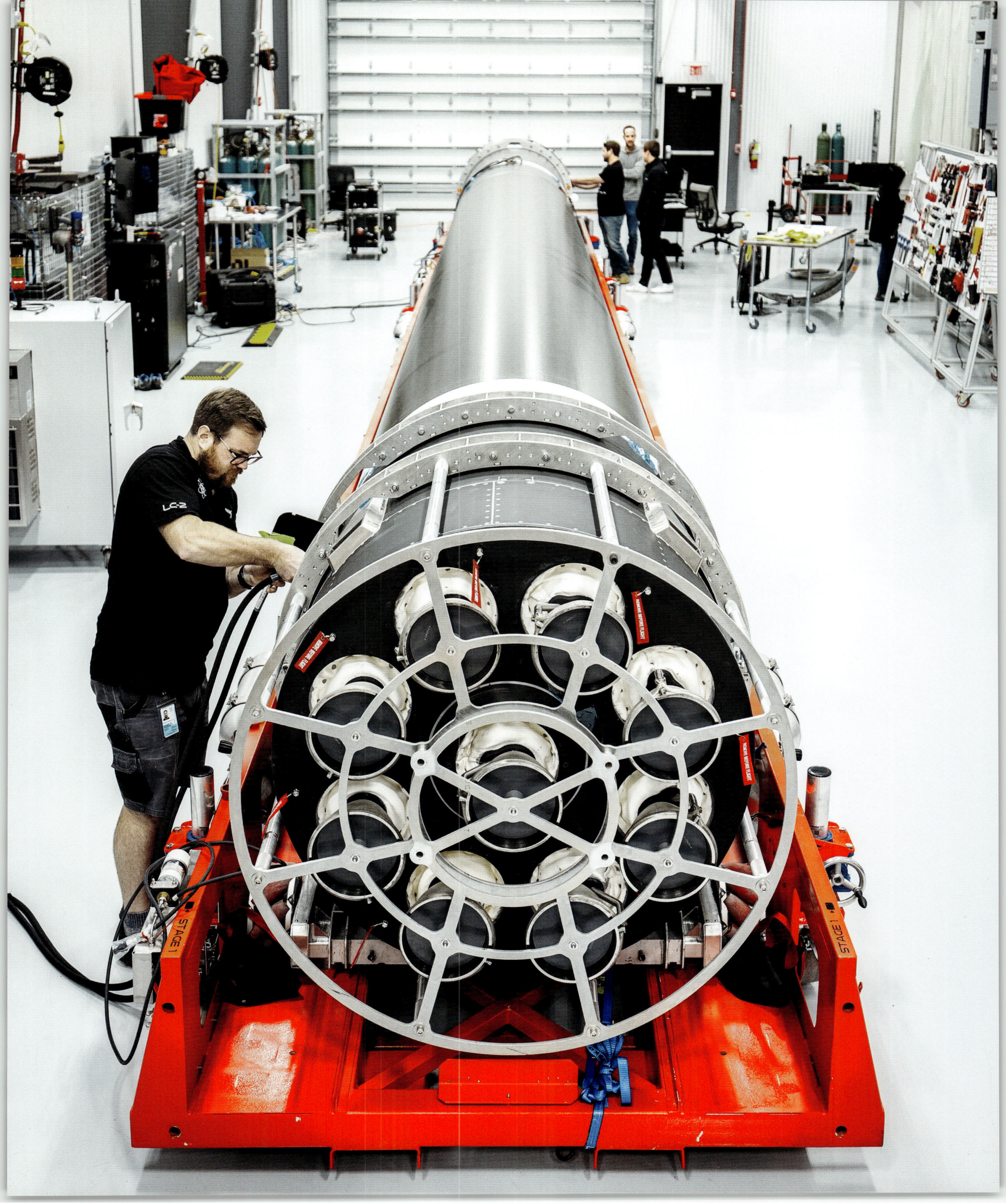
LC-2
STAGE 1
STAGE 1

3D PR

Rocket Lab's approach to 3D printing goes beyond simply adopting existing technology. The company has pushed the boundaries of what's possible with additive manufacturing.

INTING

3D Printing Adds Up

The team was simultaneously honing another innovation that would become integral to the success of Rutherford, and Rocket Lab's business case—3D printing.

"We printed geometry that even today, if you take to most 3D-printing shops, they'll tell you that's not a printable geometry. A lot of the components even now push the boundaries of what would generally be acceptable."

— Sir Peter Beck

A nascent technology in 2013, no orbital rocket engines had been developed with 3D printers, though NASA had spent years exploring the technology.

3D printing engine parts had numerous advantages. Customized engine parts are typically made through a process of subtractive manufacturing, when a big block of metal is gradually whittled away to produce the required part. Additive manufacturing using a 3D printer instead builds up layers of metal material, achieving greater precision in the process.

"3D printing is fantastic for things with internal geometries that you can't get with traditional manufacturing techniques," said Lachlan Matchett. "Flow inside the thrust chambers is a great example, where the kerosene flows throughout the wall of the thrust chamber, then into the injector. With any other manufacturing technique, it's horrendously expensive and time consuming to create. 3D printing eliminates all that."

A decade ago, Rocket Lab faced skepticism when it showed off its first 3D-printed engine. It certainly raised eyebrows at the annual Space Symposium held in Denver, Colorado in 2015, when Peter turned up to launch the Rutherford engine.

"When we started, we bought 3D printers, and we hacked into them," Peter told *MIT Technology Review* in 2019. "We modified it to suit our requirements. We printed geometry that even today, if you take to most 3D-printing shops, they'll tell you that's not a printable geometry. A lot of the components even now push the boundaries of what would generally be acceptable."

NASA undertook an in-depth analysis of the Rutherford engine as a condition of selecting Rocket Lab for a satellite-launching mission, including examining whether the 3D-printed engine with electric turbopumps and batteries would work.

"It all passed muster," said Peter.

Since then, over 600 Rutherford engines have gone to space. Now most rocket manufacturers make extensive use of 3D printing, a method which allows Rutherford and Electron components to be printed in just twenty-four hours. Rocket Lab has since demonstrated the durability of its 3D-printed engines by recovering and refurbishing them for reuse.

Right and overleaf: 3D-printed parts at Rocket Lab's headquarters in Long Beach, California, US.

Opposite: Inspection of a Stage 2 Rutherford engine thrust chamber.

Lachlan has literally grown up with Rocket Lab, progressing from Design Engineer to Propulsion Team Lead, which saw him hire some of Rocket Lab's first US-based staff, to Vice President—Propulsion, and now Deputy Chief Engineer. Hundreds of people at Rocket Lab are now involved in designing and building the company's propulsion systems, including Archimedes, the larger engine that will power the Neutron medium-lift rocket.

It's a world away from the small facility by Auckland Airport where Lachlan and team would regularly scorch the grass on the airport grounds in the process of hot-firing engines. "We didn't let anything get in our way in those early days," he said.

One engine test was disrupted when a liquid oxygen valve fitting disappeared. The team scoured the whole site, but couldn't find the crucial fitting, suspecting it had been sent back to Rocket Lab's facility in a garbage bag by accident.

"It was 8 p.m. We called Pete and told him it might be in the trash," Lachlan remembered. "So he put on his white bunny suit, got his gloves on, and hopped in the dumpster with all these black garbage bags. He started tearing through them trying to find the fitting that we needed to do that test that night."

To the embarrassment of Lachlan and his hot-firing crew, the missing valve fitting was found on the test site.

"We had our CEO rummaging through the dumpster to help get this test done. But that's how we all did it. If it's a shitty small task through to something complicated, we'd all just get together and sort it out. Don't let anything get in our way."

Pioneering Use of 3D Printers

Rocket Lab's approach to 3D printing goes beyond simply adopting existing technology. The company has pushed the boundaries of what's possible with additive manufacturing in several ways:

1

CUSTOM MODIFICATIONS

When Rocket Lab started, they purchased 3D printers and modified them to suit their specific requirements. This customization allowed them to print geometries that were considered impossible by most 3D-printing shops at the time.

2

COMPLEX COMPONENT INTEGRATION

Rather than 3D printing simple parts like brackets, Rocket Lab focused on complex components that could be merged into more efficient subsystems. This approach maximizes the benefits of additive manufacturing.

3

INJECTOR DESIGN

The company 3D prints all of its injectors, incorporating geometries that allow for superior mixing and performance, which would be impossible with traditional manufacturing processes.

4

COMBINED STRUCTURES

By 3D printing single-piece structures that join multiple components and fluidics together, Rocket Lab has found a truly useful application of the technology.

GOO GONE
PRO-POWER

3M

Above: Rocket Lab's additive manufacturing facility in Long Beach, California, US.

Previous page, left: Rocket Lab technicians inspect a newly printed rocket engine component.

Previous page, right: Powder clean-up of a 3D-printed rocket engine component.

Above: Additive manufacturing in progress.

Overleaf left: 3D-printed rocket engine quarter segments in production at Rocket Lab's Engine Development Complex in Long Beach, California, US.

Overleaf right: 3D-printed rocket engine quarter segments.

soluk
I-O
ROOF
MOVE
DATA

The Lightweights of Rocketry

When Rocket Lab launched Electron in 2017, it was the first orbital-class rocket to rely entirely on carbon composites for its primary structures.

At the time, most rockets used aluminum alloys due to their established reliability and ease of manufacturing. However, aluminum is heavier than carbon composites, limiting payload capacity and increasing costs.

Peter Beck, influenced by his research at IRL with advanced composite structures, always had his heart set on constructing a rocket from carbon fiber-based materials. Many aerospace companies had experimented with carbon composites and even launched carbon composite suborbital rockets. But the industry had given up using the material for a rocket capable of putting satellites in orbit around Earth.

By 2013, Rocket Lab had fifteen full-time employees, and Peter was determined to ramp up development of Electron's carbon composite body with exactly that goal in mind. Ben Malcolm, then a thirty-one-year-old boatbuilder and carbon composites specialist who had worked with Emirates Team New Zealand in the America's Cup yacht racing syndicate since 2007, was one of the first ones through the door.

"I was invited into Rocket Lab to do some laminating on what was going to be a composite rocket," said Ben. "When I was told that I thought, 'Whoa, in New Zealand, really? No way, this is a bit of a joke.'"

Ben grew up in Rotorua, a landlocked tourist town in New Zealand's central North Island which has a number of beautiful lakes, and is known for its geothermal activity, which produces occasionally effusive geysers. After attending a polytechnic, he landed at Salthouse Boatbuilders, one of New Zealand's most iconic marine industry firms, which by the time he arrived in the early 2000s was moving into producing high-tech carbon racing yachts.

From Yacht Building to Rockets

On arriving at Rocket Lab, he soon realized that Peter definitely wasn't joking about building a carbon fiber rocket. The plan was to construct Electron's body entirely from carbon fiber-reinforced polymers (CFRP), drawing on the techniques that had allowed Emirates Team New Zealand's designers and boatbuilders to continuously innovate in the use of the materials.

Peter also wanted to tap into the high-performance culture that went along with elite yacht racing.

These pages: Rocket Lab technicians work inside the first-stage tank of an Electron rocket.

"Electron might be sitting on the launchpad at negative 180°C. Then you're launching and you've got this friction. So you've got, like, a 250-degree delta, or difference in the temperature the materials are exposed to."

— Ben Malcolm

Above: Ben Malcolm, Senior Advisor—Composites, Rocket Lab.

Right: Electron rockets on the production line at Rocket Lab's Auckland production complex, New Zealand.

"That's what the Cup has, the prestige of this incredible thing that you achieve over three or four years," said Ben. "Our guys wanted to be a part of something again."

A number of former Emirates Team New Zealand colleagues joined him in the carbon composites team at Rocket Lab, and they set to work tackling some of the specific challenges associated with using carbon fiber to build a device capable of blasting into space and delivering satellites into orbit.

Those challenges included developing materials that could deal with the extreme temperatures associated with putting a rocket into space.

"Electron might be sitting on the launchpad at negative 180°C [356 degrees Fahrenheit]. Then you're launching and you've got this friction. So you've got, like, a 250-degree delta, or difference in the temperature the materials are exposed to," explained Ben.

Traditional carbon composite fuel tanks were prone to cracking or delamination when exposed to cryogenic temperatures. Ben's team developed a proprietary resin system and reinforcement techniques that ensured their tanks could securely hold liquid oxygen without compromising structural integrity.

The team also had to come up with new precision manufacturing techniques to build Electron's thin-walled composite structures. The walls of Electron's fuel tanks are only slightly thicker than a credit card, demanding meticulous quality control to prevent defects that could damage the tanks and explode the cryogenic liquid oxygen or the refined kerosene it reacts with to power the Rutherford engines.

Techniques like spread-tow carbon fabrics—used in racing yachts to reduce weight while maintaining stiffness—were adapted for Electron.

With Peter's goal of regular Electron launches, Rocket Lab needed a manufacturing process that could produce composite components rapidly and consistently. Later, the requirement that Electron rocket stages be reusable meant particular attention had to be paid to producing materials capable of withstanding the force of multiple launches and reentry.

Ben was used to hard deadlines and the one-hundred-hour weeks that went into meeting them—but he knew the failure tolerance was even lower in the space industry.

"There are no sea trials with a rocket," he noted. "You leave the dock, or the launchpad, and there's no turning back because you've forgotten your wet-weather gear or there's something wrong with the rudder."

The tiniest flaw in Electron's materials could see it blow up on the launchpad or in flight. Ben remembers the huge hours the team put into development, regularly sacrificing weekends, to refine and test the carbon composite materials. The effort soon began producing results.

Significantly lighter than metals like aluminum and stainless steel, use of carbon fiber reduces the overall mass of the rocket. That weight saving translates directly into increased payload capacity and lower launch costs.

The strength-to-weight ratio of the material is also a huge advantage. Despite their lightness, carbon composites are exceptionally strong and able to withstand the intense forces and vibrations experienced during launch.

But carbon composites also offered the advantage of reducing materials and construction costs, an imperative given Rocket Lab's limited funding in its early days.

Thinking Instead of Spending

"One of the biggest issues with the aerospace industry is that they just love spending money, and sometimes don't even know why the hell they're doing it," Ben said. "We were faced with trying to do things not in the Boeing way, not in the Lockheed Martin way, but the way the marine industry had done things for so many years."

That was the cheap way.

With the mindset of a home builder, Ben was tasked with producing do-it-yourself solutions to expensive problems. The composites team constructed their own industrial ovens for curing materials. If there wasn't budget for new tools, Ben and his colleagues often paid out of their own pockets, considering it an investment in Rocket Lab's future.

"Early on, if I said to Pete we'd need to have an autoclave or processing equipment, he'd be like, 'Well, can you build it? Can you make it? Is there another way of doing it? Have you thought about every way that you can do it?'"

Peter knew New Zealand boatbuilders were the masters of improvisation, mirroring his own pathway into the space industry. "I don't need a PhD in transonic aerodynamics," he told the *New Zealand Herald*. "What I actually need is aircraft technicians and more composite builders—those trades and skills are very hard to come by."

By 2017 and Electron's maiden launch, there were twenty-five people in Rocket Lab's composites team, a third of whom had worked with Team New Zealand. Electron's towering 55-foot-tall (17 m) structure stood as testament to the skill of boatbuilders turned rocket engineers.

Rocket Lab's pioneering work with carbon composites has had a lasting impact on the aerospace industry. Other companies have since adopted composite materials for rocket structures, recognizing their potential for weight savings and cost reduction.

Proven techniques further evolved by Rocket Lab—such as automated fiber placement (AFP)—have set new standards for efficiency in composite fabrication.

All of the techniques Ben and his team applied to Electron are now in play with Neutron, Rocket Lab's medium-lift reusable rocket which is also made entirely from advanced carbon composites.

Neutron is the largest carbon composite rocket ever made. In 2024, Rocket Lab's Space Structures Complex in Middle River, Maryland took delivery of a 90-ton (81 tonne) automated fiber placement machine—the largest of its kind in the world—that is custom built to accommodate Neutron's dimensions. Standing 39 feet (12 m) tall, the custom-made AFP machine is capable of laying 328 feet (100 m) of continuous carbon fiber per minute.

It will reduce the time taken to produce a stage-two dome for Neutron from weeks to twenty-four hours, allowing Rocket Lab to save over 150,000 hours a year constructing rockets.

These days, Ben divides his time between New Zealand and Newport, Rhode Island, a mecca for carbon composite development. Rocket Lab's manufacturing facilities have expanded significantly since the day he walked onto the cramped shop floor at Rocket Lab's Airpark Drive, Auckland factory, where precision electronics were assembled alongside builders grinding carbon.

He remembers riding his skateboard across the concrete floor of the empty new factory at Levene Place in Mount Wellington, a suburb in the east of Auckland, where Electron rockets undergo final integration ahead of delivery to LC-1 at Mahia.

The first Electron prototypes featured rocket stages a handful of people could pick up.

"Stage two was the length of two computer desks, enough for a 33 to 44-pound (15-to-30 kg) payload. It just kept changing as people became interested in the company. We'd always finish something, and then we'd have to make it bigger," he said.

Neutron continues this theme. With a height of 141 feet (43 m), and 23 feet (7 m) in diameter, even finding a factory big enough to accommodate it is a challenge.

"You're building foundations for skyscrapers to test these things," Ben said. "They just have so much weight inside of them, it's insane."

Ben sees the ongoing relationship between the marine industry and Rocket Lab as more important than ever before, and one that has helped sustain yacht builders who are often at the mercy of exchange rates and fickle superyacht owners.

"Stage one was the length of two computer desks, enough for a 33 to 44-pound (15-to-30 kg) payload. It just kept changing as people became interested in the company. We'd always finish something, and then we'd have to make it bigger."

— Ben Malcolm

Right: An Electron rocket in production at Rocket Lab's Auckland production complex, New Zealand.

REMOVE BEFORE FLIGHT
ROCKETLAB
REMOVE BEFORE FLIGHT
REMOVE BEFORE FLIGHT
REMOVE BEFORE FLIGHT
REMOVE BEFORE FLIGHT

MA

HIA

Mahia: Rocket Lab's Gateway to Space

With funding in hand and Electron taking shape, Peter Beck's attention turned to an important practical consideration—where in the world to launch his rockets from.

This choice not only shaped Rocket Lab's trajectory, but also transformed a quiet coastal region of New Zealand into a bustling spaceport, and marked the country's entry into the global space industry.

The quest for an ideal launch location began in earnest in 2015. Peter had a clear set of criteria. Primarily, he needed a location that offered a wide range of orbital inclinations, minimal air and marine traffic, and the ability to eventually support a weekly launch schedule.

He was adamant that Rocket Lab needed its own private launch facility. The company's business model rested on it. Using an existing launch facility at a NASA site or US spaceport can come with a price tag of up to $2 million per launch.

"Turning up to the United States to try and launch off Cape Canaveral, there was no way we could have afforded to do that. It would have blown the business case completely apart," said Peter.

He gave Shaun D'Mello, another young engineer who had joined Rocket Lab straight out of university, the task of finding a suitable launch site—and developing the launch infrastructure.

Shaun, who now serves as Vice President—Neutron at Rocket Lab, grew up with a passion for aviation, and had intended to become a pilot. But he soon became more interested in engineering than flying aircraft.

Whilst living in Australia, he enrolled in a Bachelor of Engineering degree at the University of Sydney. It was there, towards the end of his studies, that he came across Rocket Lab.

With little in the way of a commercial aerospace industry in Australia at the time, he submitted his resumé via Rocket Lab's website in the hope of scoring an internship. "I added a cover letter that basically said, 'I'm willing to do whatever to help,'" he said.

He soon found himself joining Rocket Lab's fourteen-strong team in Auckland, and just a few months later was taking helicopter rides with Peter around the country to scout out potential launch sites.

His unfamiliarity with New Zealand had advantages and disadvantages. He had no preconceived ideas about where to look for the perfect launch site, but was also prone to getting excited about locations that happened to be in the middle of the country's sprawling national parks, which are off-limits to most kinds of commercial activity.

An obvious candidate for the launch site was Kaitorete Spit, a long and isolated finger of land stretching along the coast of Canterbury in New Zealand's South Island. Indeed, Rocket Lab would later announce Birdlings Flat on the eastern end of the spit as its preferred launch site.

It was close to Christchurch, the country's second-largest city, with good international air connections and strong engineering capability in the Canterbury region. Peter envisaged building a manufacturing facility in Christchurch to assemble the Electron rockets for transfer to the spit.

But Kaitorete also had some big drawbacks. While suitable for Sun-Synchronous Orbital flights, it would not permit mid-inclination launches due to the presence of Banks Peninsula to the east. Rocket Lab also encountered delays in securing resource consents from the local authorities. Crucially, the consents also permitted Rocket Lab to undertake just one launch per month, well behind Peter's desired launch cadence of one per week, or more, once Rocket Lab had proven Electron and scaled up production.

Māori have occupied the peninsula for hundreds of years, naming it Te Mahia, te reo Māori for "indistinct sound." For seventy years, the farmland had been held on behalf of the local Māori, and run as a large sheep farm.

Above and previous page: Mahia Peninsula, New Zealand, before Rocket Lab Launch Complex 1 was built.

"I was just stunned by the landscape. It felt very surreal to be on this hilltop looking down the peninsula and seeing the very furthest point, and water all around. It was like the textbook launch site, because it had everything you look for . . ."

— Shaun D'Mello

Shaun D'Mello, Vice President—Neutron, Rocket Lab.

A Launch Path Over the Pacific

Shaun and Peter continued to look to New Zealand's isolated east coast for an alternative site. A stretch of land jutting out southward into the sea between Poverty Bay and Hawke's Bay, two-thirds of the way down New Zealand's North Island, caught their attention.

It was far from any cities, awkward to get to, and occupied by sheep farms. Logistically, it was problematic. But one afternoon in late 2015, Shaun found himself at the end of a dusty gravel road at the southern tip of the Mahia Peninsula.

"I was just stunned by the landscape," he recalled. "It felt very surreal to be on this hilltop looking down the peninsula and seeing the very furthest point, and water all around. It was like the textbook launch site, because it had everything you look for—all the trajectories we analyzed in the office could safely fly from this very launch point."

Mahia Peninsula had an unobstructed launch path over the Pacific Ocean. The land Shaun stood on formed a raised plateau with an altitude of 341 feet (104 m), providing a natural platform for launch. The peninsula was positioned in fairly quiet airspace, and marine traffic consisted mainly of fishing boats and cargo ships and tankers shuttling up and down the coast.

But Mahia also had its challenges. Around 18.5 miles (30 km) of gravel road would have to be sealed to allow Electron and equipment to be delivered to what would become Rocket Lab Launch Complex 1. Most importantly, permission needed to be secured from local landowners and regional authorities.

Luckily, an old ally of Peter's again stepped up to smooth the way for launch. Sir Michael Fay, who had given over use of Great Mercury Island to Peter for the launch of Ātea-1, and graciously wined and dined visiting US aerospace representatives on his behalf, owned a farm on the Mahia Peninsula.

Michael knew the owners of Onenui Station, the farmland Shaun had identified as the perfect launch site. In February 2015, he made a phone call to Ben Mackey, the chair of the Onenui management committee, which represented local Māori who owned the southern tip of the Mahia Peninsula.

A Partnership with Māori

Māori have occupied the peninsula for hundreds of years, naming it Te Mahia, te reo Māori for "indistinct sound." For seventy years, the farmland had been held on behalf of the local Māori and run as a large sheep farm.

The call led to a meeting in Napier, an east coast city at the southern end of Hawke's Bay, south of Mahia. Peter and Shaun joined Michael at Mister D, a popular Napier lunch spot, and waited for Ben Mackey and his son George to arrive.

"Michael was the drawcard to get everybody along, so that people thought we were serious," said Peter, who was nervous about the meeting. "How do you go about turning up and saying, 'I want to launch a rocket off your farm?'" he said.

There was a lot riding on the meeting. But for the Mackeys, Peter's request represented a rare opportunity to explore a new avenue for making productive use of farmland at a time when wool prices were depressed.

Peter employed the same open and forthright approach that had worked with Silicon Valley's most seasoned venture capitalists. But this time he didn't expect to get an answer within half an hour.

The Māori of Aotearoa New Zealand value relationships and trust above all. Te ao Māori is a holistic worldview that places great importance on guardianship of the natural environment (kaitiakitanga), an approach to decision-making that focuses on respect, fairness, and consensus building (tikanga), and community and collective responsibility (manaakitanga).

Peter had grown accustomed to the transactional nature of deal-making with aerospace companies. But working with the Mahia community and landowners would require a deeper level of engagement, one that would see him need to satisfy the people of Mahia that he had their best interests at heart.

George Mackey, who would later take over from his father as chair of the Tawapata South management committee, which owns the Onenui farm, was under no illusions about how unconventional Peter's proposal was.

"We looked at each other and thought, what the heck just happened here?" he would recount to the government website New Zealand Story. "It was really quite surreal. There was an opportunity to support a fellow Kiwi who is doing amazing things."

Consultation with iwi shareholders followed, culminating in September 2015 in a meeting at Kaiuku Marae (a traditional Māori tribal meeting place) where representatives of the 1,700 shareholders in the sheep-farming station met Peter and peppered him with questions.

Again, his passion and vision proved decisive. Mackey saw a striking connection between what Peter was setting out to do, and the uncertain voyage his ancestors had made through the Pacific to Aotearoa New Zealand over a thousand years ago, using the stars to guide them.

"We also talked about the connection we've had with the stars and the gods, and [how] we are going back to visit them," he recalled.

Opposite top and left: Construction begins on Rocket Lab Launch Complex 1.

"That was one of the things that clinched the deal with our shareholders."

Once Tawapata South Inc. and local iwi gave their blessing for Rocket Lab to proceed with the Mahia site, events moved quickly. Rocket Lab filed for resource consent, which was approved by the Wairoa District Council and the Hawke's Bay Regional Council.

The Mahia site was granted a license for launches every seventy-two hours over thirty years, accommodating Peter's ambitious plans to scale up launch activity. Wairoa's long-serving mayor, Craig Little, became an enthusiastic supporter of Rocket Lab's Mahia launch development, recognizing the opportunities for economic development.

Shaun found himself in deeply unfamiliar territory, shifting out of engineer mode to negotiate lease agreements, consult with the local community, bone up on launch safety protocols, and figure out how to set up launch operations in remote Mahia. At the same time, he had to assemble an engineering team to develop the infrastructure for LC-1, and think about operations for actually launching Electron.

Rocket Lab broke ground at Mahia in December 2015, and by June 2016, most of the infrastructure work had been completed. This included upgrades to roads and internet services, and the development of electrical, mechanical, and communications infrastructure.

"At that time, the amount of infrastructure that we had to build to make a viable launch site was just crazy," said Peter. "It used to take us forty-five minutes to get from the farmhouse down to the launch site in a four-wheel drive. Now it takes five minutes on a sealed road."

But he has no regrets about choosing the Mahia Peninsula, which has played host to more than sixty Electron launches to date.

Shaun takes pride in the fact that many years later, LC-1 is largely unchanged, apart from the addition of a second launchpad, and an upgraded range control building.

Rocket Lab LC-1 was officially opened on 26 September 2016, in a ceremony attended by company employees, local landowners, and government officials. It was hosted by the New Zealand government's then-economic development minister, Steven Joyce, who had worked solidly during the year to create the complex regulatory framework to allow for New Zealand to host rocket launches.

There wasn't much time to celebrate, as Peter's attention turned to Electron's looming maiden launch. But as he drove back along the freshly sealed road winding through the fields, he felt vindicated at the decision to base Rocket Lab's ambitions on this breathtaking, isolated, and challenging launch site.

"Sometimes the hardest path is the best path," he reflected. "Mahia was definitely the hardest path."

Below left: The strongback arrives at Launch Complex 1 for the Electron launch platform on Pad A.

Below right: Launch Complex 1 Pad A.

Opposite: Successful lift-off of Rocket Lab's eighteenth Electron launch, "Another One Leaves The Crust."

Overleaf: Mahia Peninsula, New Zealand, after construction of Rocket Lab Launch Complex 1.

"Sometimes the hardest path is the best path. Mahia was definitely the hardest path."

— Sir Peter Beck

LAUNCH COMPLEX:

1

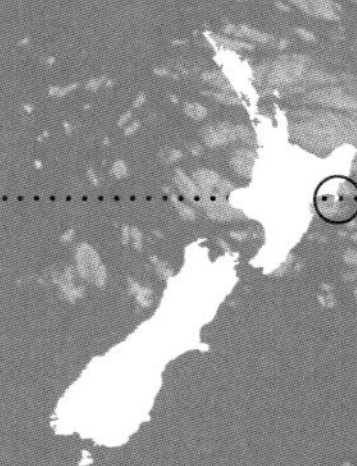

Mahia Peninsula,
New Zealand

LONG 39.26085°S
LAT 177.86586°E

Rocket Lab's first and primary launch site, Launch Complex 1, is located on the picturesque Mahia Peninsula on the east coast of New Zealand's North Island. Opened in 2016, the facility has the distinction of being the world's first private orbital launch complex.

INFRASTRUCTURE AND CAPABILITIES

Launch Complex 1 boasts impressive infrastructure designed to support a high launch cadence:

- Two launch pads for increased operational flexibility
- A vehicle integration facility capable of processing multiple Electron rockets simultaneously
- A state-of-the-art range control center for monitoring and controlling launches

The site's remote location offers several strategic advantages:

- Minimal air and marine traffic, allowing for frequent launches
- A wide range of achievable orbital inclinations
- An unobstructed launch path over the ocean, enhancing safety

Rocket Lab added a second launch pad, named Launch Complex 1B (LC-1B), to their Mahia Peninsula facility in New Zealand in February 2022.

Construction of LC-1B began in December 2019, and it officially hosted its first launch on February 28, 2022, with the mission "The Owl's Night Continues." This new pad is located 383 feet (117 m) from the original Launch Complex 1A (LC-1A). The addition of LC-1B effectively doubled Rocket Lab's launch capacity at the Mahia Peninsula site, providing greater flexibility and enabling a higher frequency of launches.

MISSION TEAM

ROCKET·LAB

We Need to Become a Launch State

By late 2015, Electron was literally coming together. Test firing of the Rutherford engines was yielding promising results.

Refinements had been made to the launch vehicle design and the various rocket stages. Work was well underway on construction of LC-1 on the Mahia Peninsula.

But Peter Beck now faced another challenge. To pave the way for commercial space activities in New Zealand, the government would have to implement substantial legislative changes given that there were no existing laws controlling launches to space from the country.

Rocket Lab also had to navigate international agreements and regulations that tightly controlled the use and transfer of technology that could be used to launch objects into outer space. Regulatory processes rarely move at a pace that accommodates ambitious startup founders. A lot of red tape now threatened to bog down the team and leave Electron gathering dust.

"Sometimes the worst thing you can do is write down all the things that you have to do. Because if you do that you would never start," said Peter, who had deliberately focused on the massive technical challenges of getting Electron ready for launch and the Mahia site developed, leaving it late in the day to gain the necessary authorizations.

Now he had to shift gears, prioritizing the bureaucratic paperwork he needed to get rubber-stamped to allow Electron to take flight. It certainly helped enormously that Rocket Lab was now a US-headquartered company. But it wasn't just a matter of shipping Rutherford engines from California to Rocket Lab's New Zealand premises.

The US and New Zealand are both signatories to the Missile Technology Control Regime (MTCR), which was established in 1987 by the G7 nations with the aim of preventing the proliferation of weapons of mass destruction by controlling the export of technologies that could form delivery systems for such weapons. That includes rockets and drones.

The MTCR is not a legally binding treaty, but its thirty-five members must jump through numerous hoops to share rocket technology with each other, let alone non-member nations.

"If any one of those thirty-five signatories refuses to grant permission, then you're not allowed to create launch capability. The reality is, if you can put a satellite in orbit, you can put a thermonuclear warhead in place of the satellite and launch it anywhere in the world," said Peter.

The US government itself had for decades denied requests from other MTCR nations seeking to develop rocket launch capabilities. Now, for the first time, it was seeking to transfer technology to another country, and support it becoming a rocket-launching nation.

The cornerstone of Rocket Lab's ability to operate in New Zealand was the Technology Safeguards Agreement (TSA), which was negotiated between New Zealand and the US and allowed Rocket Lab's activities in New Zealand to comply with the MTCR.

Help with securing the agreement was top of Peter's list of requests when Steven Joyce paid a visit in late 2015 to Rocket Lab's assembly plant next to Auckland Airport.

"He had some specific asks, and obviously the regulatory ask was the key one," Steven remembered.

A former broadcasting entrepreneur and the son of grocers, Steven held the economic development, and science and innovation portfolios in then-Prime Minister John Key's National Party-led coalition government. He had a reputation as National's "Mr. Fix-it," able to get complex and politically challenging projects

Previous page and overleaf: An Electron rocket, Launch Complex 1, Mahia Peninsula, New Zealand.

over the line, and was the driving force behind a successful government-backed scheme to build a fiber-optic broadband network throughout New Zealand.

He knew about Rocket Lab, which by 2016 had been the recipient of government R&D grants, and was a darling of the startup scene. But Peter's requests didn't seem to account for the often-glacial pace of bilateral deal-brokering.

"He sort of reeled off a bunch of stuff, and said he needed it in six months," Steven said. "I thought, 'Ooh, that'll be challenging.'"

Getting the Government Onboard

But his passion for innovation and his government's desire to diversify the New Zealand economy had Joyce wondering just how to help Peter meet his deadline, as he flew back to Wellington.

The visit to the Rocket Lab factory was transformative for Joyce, who was used to getting pitches from entrepreneurs convinced their company could put New Zealand on the map and make billions for the country.

"I just remember walking around the factory there going, 'Wow,'" he said. "It was Peter's intensity, his knowledge of things, and his drive and determination which was immediately apparent, and sort of lifted him well above the rest of the pantheon of inventors around the place."

Back in the capital's beehive-shaped government building, Steven started putting a team together, drawn largely from two key government departments, the Ministry for Business, Innovation and Employment, and the Ministry of Foreign Affairs and Trade.

"There was a lot of sucking of teeth in Wellington about whether this was possible or not," he said.

But the plan had the full backing of Prime Minister John Key, who also happened to have developed a close relationship with then-President Barack Obama, forged as the pair met at diplomatic summits around the world, and spent time together on the golf course in Hawai'i, where they both had holiday homes.

That high-level relationship helped accelerate the timeline. But the real key to getting the TSA approved on the other side by the US government was the strength of the support for it that came from within the US defense agencies.

After all, it was the Operationally Responsive Space Office, based at the US Department of Defense, that in 2010 awarded Rocket Lab a contract to develop a low-cost space launcher for putting CubeSats into orbit.

The Office of Naval Research, responsible for the US Navy Marine Corps science and technology programs, and the Defense Advanced Research Projects Agency had also funded Rocket Lab projects, with the specific aim of developing small launch capability—something NASA's own rocket program and even SpaceX weren't delivering.

"There was a big meeting at the Department of State where we had all these people from different agencies saying, 'We need this,'" said Peter.

A key advocate in the Department of State was Rose Gottemoeller, who as Under Secretary for Arms Control and International Security was the government's key adviser on arms control and nonproliferation. Rose would go on to serve as Deputy Secretary General of NATO from October 2016 to October 2019.

The support at multiple levels in the US government cleared the way for the TSA to be signed in June 2016. It enabled the use and secure management of US rocket and satellite technology in New Zealand, and allowed Rocket Lab to import launch technology from the US. It was the first TSA to be signed between the US and another party to the MTCR. The US has subsequently signed agreements with the United Kingdom, Brazil, and Australia.

That agreement satisfied the Americans. But Steven also had to shepherd through the creation of a legislative framework to govern space launches and satellite operations from New Zealand, and to give effect to the various international treaty provisions it was obliged to observe.

In December 2015, John Key's Cabinet approved the high-level design of a regulatory regime. It led to the creation of the Outer Space and High-altitude Activities Act 2017 (OSHAA), and the establishment of the New Zealand Space Agency.

> "There's a halo effect, which is apparent. It's not just in the space industry, but a broader sense of, 'We can do anything from here.' There's only one reason New Zealand has a space industry, and that's because of Peter Beck."
>
> – Steven Joyce—former New Zealand government Minister for Economic Development

ELEC

TRON

Right: Looking out to sea from Mahia Peninsula, New Zealand.

The Ultimate Test—Electron's Maiden Flight

May 25, 2017 stands out as a pivotal moment not only for both Rocket Lab, but the commercial space industry in general.

Up until that point, SpaceX was the only private rocket company regularly launching small satellites into Low Earth Orbit (LEO). Elon Musk had outsmarted the incumbent aerospace giants to win billions of dollars' worth of contracts from NASA and other US government agencies, and commercial satellite operators. The billionaire tech visionary had built a lucrative monopoly on commercial launches in the process.

That situation was about to change. Rocket Lab had spent four years developing Electron, a revolutionary small satellite launcher designed to provide dedicated, cost-effective access to orbit. The company faced the daunting task of not only creating a new rocket, but also establishing the world's first fully commercial privately operated orbital-class launch range. In contrast, SpaceX relied on US government launch facilities to get its rockets into space.

After hundreds of hot-fire tests of the Rutherford engines, stress-testing of Electron stage components, and tweaking of electronics, software and controller systems, the Rocket Lab team was ready to prepare for launch.

In February 2017, the first Electron rocket was transported to its launch base on the Mahia Peninsula. The following months were dedicated to commissioning and testing the ground systems crucial for fueling the launcher, tracking its flight, and ensuring safety protocols were in place.

How do you simulate the complex processes involved in a rocket launch?

"A lot of role-playing," said Michael Pearson, the then-Propulsion Engineer, now Director—Test and Launch Operations, tasked with overseeing many of those final tests. "It's like Dungeons & Dragons, but with rockets instead," he said.

The team had software tools to simulate the rocket's behavior and the ground systems controlling it. Like pilots sitting in a flight simulator, the launch team was confronted with errors and system failures to test their ability to react and make decisions in real time.

In the weeks before launch, the testing and practice involved real hardware. Electron sat on the pad in Mahia as the team practiced filling it with propellants, and every aspect of the launch countdown was rehearsed.

"We've spent probably the last month [doing] a lot of conditioning and fault-finding on all the ground systems," Peter Beck told *Spaceflight Now*. "People kind of see these launch towers as lumps of steel and whatnot and think they're relatively benign, but boy, there's a lot going on there."

As with most pivotal stages of Rocket Lab's history, it was a period marked with frenetic work, sixteen-plus-hour work days.

"I think I've blocked most of it out of my mind," said Michael. "It was traumatic, but it was very good. I remember not sleeping very well for a week or two in the lead-up, not just because I was nervous, but also because there was so much work to get done."

"It was exciting," remembered the unflappable Shaun O'Donnell, already a ten-year veteran of Rocket Lab by that stage. "There was a huge amount of different things going on all at once, and you could see anticipation building in the engineers, getting towards that final point when they could prove a lot of what they've been working on for years."

"It's a Test"

This maiden flight was aptly named "It's a Test." It didn't carry any commercial satellites or operational payloads. Instead, Electron was fitted with instrumentation that would collect 25,000 channels of telemetry for engineers to evaluate the launcher's performance. The mission aimed to reach an eighty-three-degree elliptical orbit of 186 by 310 miles (300 by 500 km).

"The first launch, in some respects, is easier from a public perception standpoint than the fiftieth launch, because you get a free pass for your first launch in rocketry," said Rocket Lab's Vice President—Marketing and Communications Morgan Connaughton, who was then a senior staffer at public relations firm Porter Novelli, responsible for the Rocket Lab account.

But Peter knew a catastrophic failure that revealed major design flaws in Electron could

"I gave them fifteen seconds, and asked if they were ready . . . no answer. I need an answer now . . . . We are a go."

— Dr. Sandy Tirtey, Senior Director—Global Commercial Launch Services, Rocket Lab

At 4.20 p.m. local time (12.20 a.m. EDT), the 55-foot-tall (17 m) Electron rocket, powered by its nine Rutherford engines, roared to life on the launchpad.

Above left: Michael Maurer, Launch Director for Electron's "It's a Test" mission.

This page: Tense moments and celebration during the launch of Rocket Lab's first Electron mission, "It's a Test."

Opposite: Dr. Sandy Tirtey, Senior Director—Global Commercial Launch Services, Rocket Lab and Peter Beck.

have been potentially fatal for the company. At the time, Rocket Lab had one completed rocket on the launchpad, and three in the factory in various stages of assembly. "If we'd blown it up on the pad, then that would have been the end," he reflected.

As the launch window approached, Mother Nature threw a curveball. Poor weather kept the Electron grounded for three consecutive days, testing the patience of the launch team, but providing valuable practice in countdown and scrub procedures.

Lachlan Matchett, then Vice President—Propulsion, had to scramble at the last minute to deal with a broken radar unit at the launch site. Luckily, one of Rocket Lab's investors had access to a helicopter. They flew from Auckland to the launch site with spare equipment to avoid the launch being delayed.

Finally, on May 25 the skies cleared, and the stage was set for Rocket Lab's moment of truth.

The Launch Conductor that day was Dr. Sandy Tirtey, the man who had been involved in virtually every aspect of Electron's development in the preceding years. At Mission Control in Auckland, he paced around wearing his wireless headset. "I'd sometimes take it into the bathroom, just to keep listening to what was going on," he said. "If I said something from the bathroom, they would all know, because the sound was a bit different."

It is often customary in the space industry for the Launch Director to voice the final ten-second countdown. But Tirtey wasn't having a bar of that. He had written some of the launch software, and wanted to be free to monitor its performance.

"I had organized a kind of robotic voice to do the countdown," he said. "Pete hated it, so he decided to do the countdown himself. He did that for like, the first ten flights."

But there would be drama before Peter could get to the T-minus ten-second countdown. Earlier, at the T-minus twelve-minute mark, Sandy heard over his headset that the range controller at Mahia had detected a leak in the igniter fluid system on the launch pad. Without enough pressure in these igniter-fluid tanks, the rocket engines could fail to ignite. They needed to get a handle on the extent of the leak—and fast.

The minutes counted down as Sandy listened eagerly for an update, while voices on the radio channel calmly confirmed that other systems were ready for launch. At five minutes, the news was concerning.

"There's a leak. We don't know by how much," was the update from Shaun D'Mello, who was leading the launch control team in Mahia. Working frantically in the background, Shaun's team was trying to calculate the leak rate, before needing to over-pressurize the system to compensate for the leak. The system was on its way to leaking down perfectly within bounds for launch, but Sandy was impatient and breathing down their necks.

At T-minus three minutes, Sandy asked the range for an update, and received an ominous silence while the leak calculations and discussions off-headset were underway.

At T-minus two minutes, he had to approve whether or not the launch could proceed. "I gave them fifteen seconds, and asked if they were ready. No answer," he said.

The seconds stretched out. "I need an answer now," Sandy said into his microphone at the ten-second mark.

He could hear Shaun breathing into his headset. Finally an answer came. "We are go."

We Were Just Cruising Home

Sandy gave the green light for launch. All eyes were glued to computer screens and the camera feeds from Mahia as the clock finally heard T-minus zero.

At 4.20 p.m. local time (12.20 a.m. EDT), the 55-foot-tall (17 m) Electron rocket, powered by its nine Rutherford engines, roared to life on the launchpad. As it blasted off, Peter and team at Mission Control leapt to their feet in elation. Electron hadn't blown up on the launchpad. The nightmare scenario had been averted.

The initial phases of the launch appeared to proceed as planned. Electron successfully passed through "max q," the point of maximum aerodynamic stress and the ultimate test of its carbon fiber shell.

The first stage performed its burn flawlessly, followed by a clean stage separation. The second stage ignited as designed, a common failure point for new rockets, and the payload fairing separated successfully. For several minutes, it seemed that Rocket Lab might achieve the rare feat of reaching orbit on its very first attempt—a distinction typically reserved for national space programs, but never before achieved by a private company.

Around four minutes into the flight, at an altitude of 139 miles (224 km), Rocket Lab's launch team could still see Electron heading for orbit on their instruments. But an independent team of contracted range safety officials monitoring the launch as a requirement of Federal Aviation Administration (FAA) regulations could not. These contractors terminated the flight, ending the mission.

Electron was terminated by simply turning off the engines, sending the rocket arching back towards Earth and burning up in the atmosphere.

"We were a couple of hundred seconds away from orbit," Peter reflected. "All of the high-risk activities were done. First stage flawless, second stage separation, fairing separation, we were just cruising home."

The frustration in Mission Control was palpable. But the team had performed its tasks with the utmost professionalism.

The reason for the termination was uncovered relatively quickly, but wouldn't be publicly reported until nearly three months later, after a thorough review had been conducted. The root cause was identified as a telemetry glitch in ground equipment, specifically a misconfiguration in a piece of equipment provided by the third-party safety contractor.

Peter said it amounted to a misconfigured software setting, with forward error correction, a method for obtaining error control in data transmissions, not being enabled in the contractor's software.

"I mean, full credit to the team. If it wasn't for one third party not ticking a tick box on a piece of software, I'm absolutely confident that thing would have made it to orbit," he said.

He drew a valuable lesson from the episode. "These contractors launch rockets all the time, and we didn't really check their work. It very much drove home a trust-but-verify culture within the company," he concluded.

"I was so pissed off that it took a human to stop it," Sandy reflected. "It was annoying that it was the professionals that had let us down. But you know, things happen. It could have happened to us, too."

The overall mood at Rocket Lab following the launch was one of relief, said Morgan. The team's hard work had been validated, with the telemetry data gathered from the rocket throughout its short flight providing a goldmine of feedback on Electron's performance.

"Everyone was exhausted, excited, proud, scared, everything under the Sun," said Morgan. "No matter what the result was, we were going to learn from it."

The investigation into the telemetry glitch found it was a relatively simple issue to address. No major changes to the Electron launch vehicle hardware were required. Procedures were updated to ensure third-party software was adequately tested ahead of launch. The results of the launch termination investigation were duly delivered to the FAA.

It was clear that the maiden launch had positioned Rocket Lab favorably for its future endeavors. Peter emphasized that reaching space on the first test flight put the company "in an incredibly strong position to accelerate the commercial phase of our program, deliver our customers to orbit, and make space open for business."

This launch demonstrated the viability of Rocket Lab's innovative approach to rocket design and manufacturing. The Electron rocket, with its electric pump-fed engines and extensive use of 3D printing, represented a new paradigm in launch vehicle technology.

With a roster of satellite customers signed up, including NASA, and Earth-observation companies Spire and Planet, and a wealth of data to sift through, the team had little time to pause and reflect before preparing for the next launch from Mahia, which would come before the year was out.

"Everyone was exhausted, excited, proud, scared, everything under the Sun. No matter what the result was, we were going to learn from it."

— Morgan Connaughton, Vice President—
Marketing & Communications, Rocket Lab

Opposite and pages 122-23:
An Electron rocket launches.

ELECTRON

ELECTRON

Rocket Lab's Electron rocket is a pioneering small-lift launch vehicle designed to provide dedicated, responsive access to orbit for small satellites.

GENERAL SPECIFICATIONS

Height:	59 feet (18 m)—but 55 feet (17 m) at time of "It's a Test" flight
Diameter:	3.9 feet (1.2 m)
Stages:	2 and Kick Stage
Launch Mass:	28,660 pounds (13,000 kg)
Propellants:	Liquid oxygen (LOX) and rocket-grade kerosene (RP-1)
Structure:	Fully carbon fiber composite monocoque (unique for orbital rockets)

PROPULSION

First Stage

Engines:	9 Rutherford sea-level engines
Thrust:	43,000 pound-force (190 kN) at lift-off, peaking at 50,600 pound-force (225 kN)
Specific Impulse (Isp):	311 seconds
Propellant Mass:	~22,487 pounds (~10,200 kg)

Second Stage

Engine:	1 vacuum-optimized Rutherford engine
Thrust:	5,800 pound-force (25.8 kN)
Isp:	343 seconds
Propellant Mass:	~4,409 pounds (~2,000 kg)

PAYLOAD CAPACITY

Low Earth Orbit (LEO):	661 pounds (300 kg)
Sun-Synchronous Orbit (SSO):	440 to 1,110 pounds (200 to 500 kg)
Payload Fairing:	4-foot (1.2 m) diameter by 8.2-foot (2.5 m) length

KICK STAGE

Electron's optional third stage enables precision orbital deployments:	Multi-orbit deployments in a single mission
	Hosted payload support (up to 374 pounds or 170 kg)
	Deorbiting capabilities to reduce space debris
Pointing Accuracy:	Fifty arcseconds

Rutherford Engine Innovations

First 3D-printed oxygen/hydrocarbon engine (Inconel alloy)

Electric turbopumps powered by lithium-polymer batteries (no gas generator)

90% efficiency with simplified design

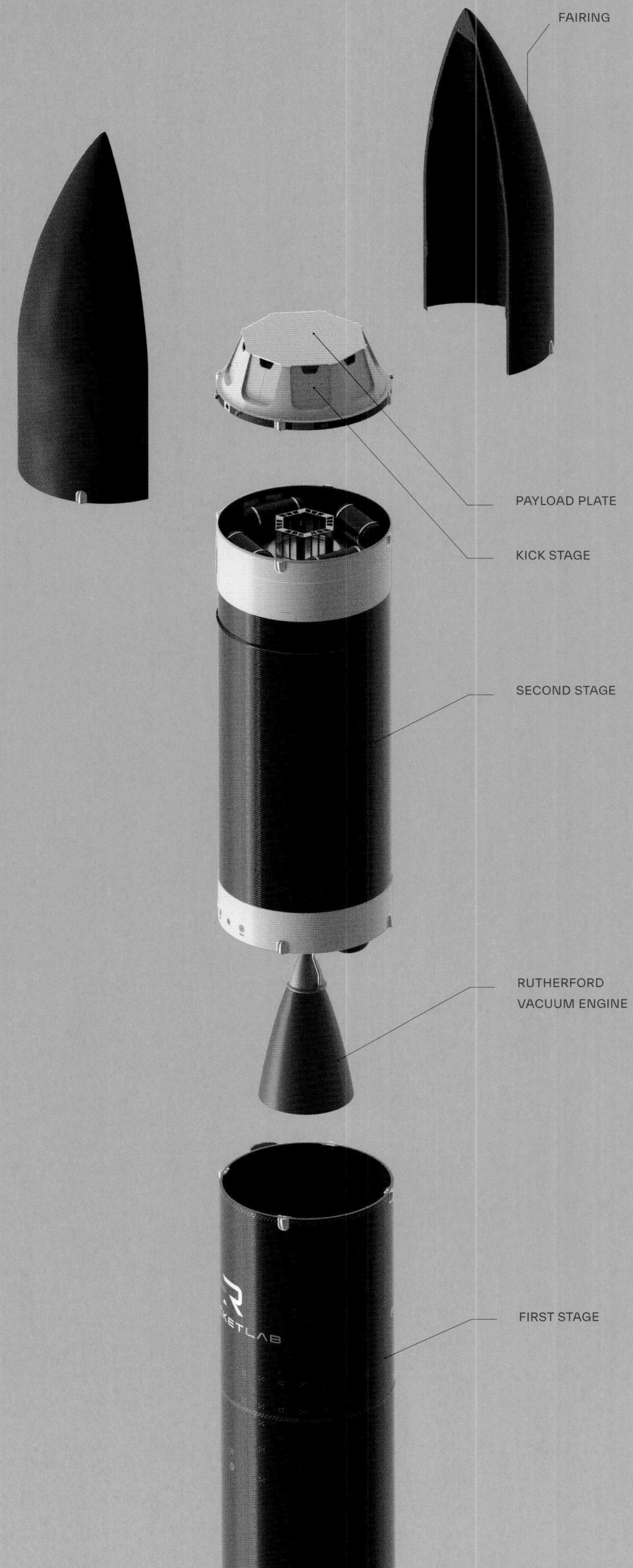

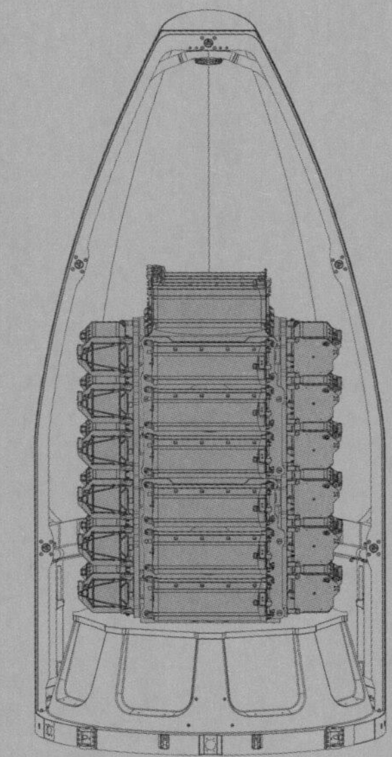

27 CUBESATS

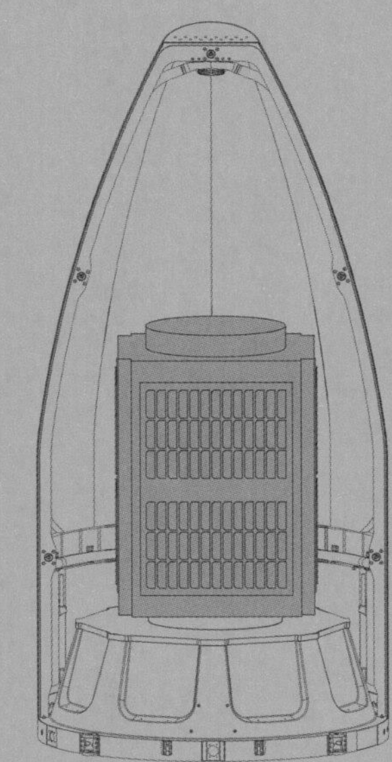

1 MICROSAT

PAYLOAD CONFIGURATIONS

Electron can support a wide range of payload configurations, from a single primary microsatellite, through to multi-satellite missions comprised of several microsats and CubeSats. For multi-satellite missions, Rocket Lab offers custom configurations for secondary payload adapters.

Pages 124-29: Electron rockets in production at Rocket Lab's Auckland production complex, New Zealand.

Page 130: An Electron rocket engine ready for launch.

Page 131: An Electron rocket blasts off.

ELECTRON

PAINT PREP
SOUND PROOF 02
RESIN MIXING BAY
F.O.D

RBF
RBF
RBF

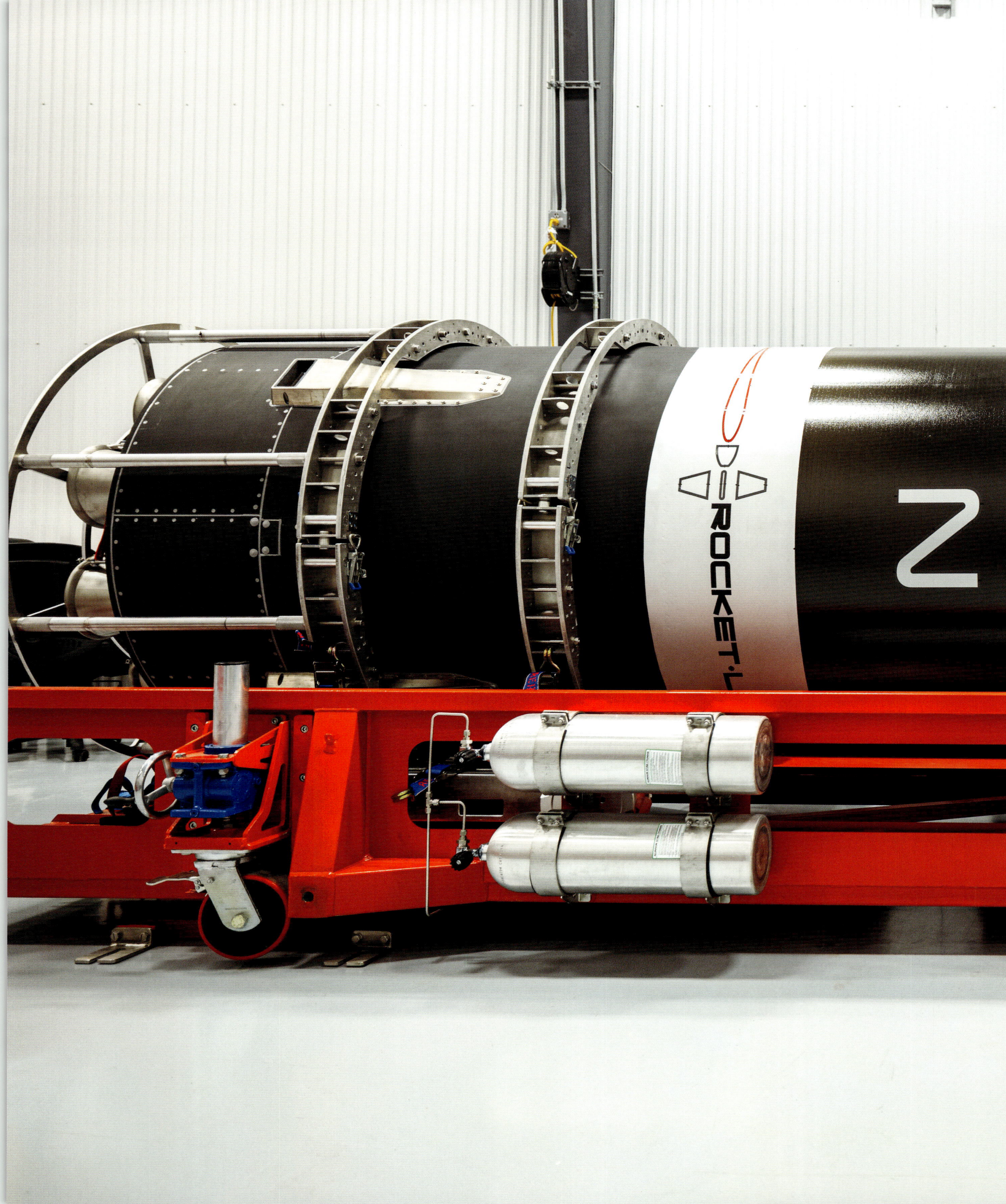

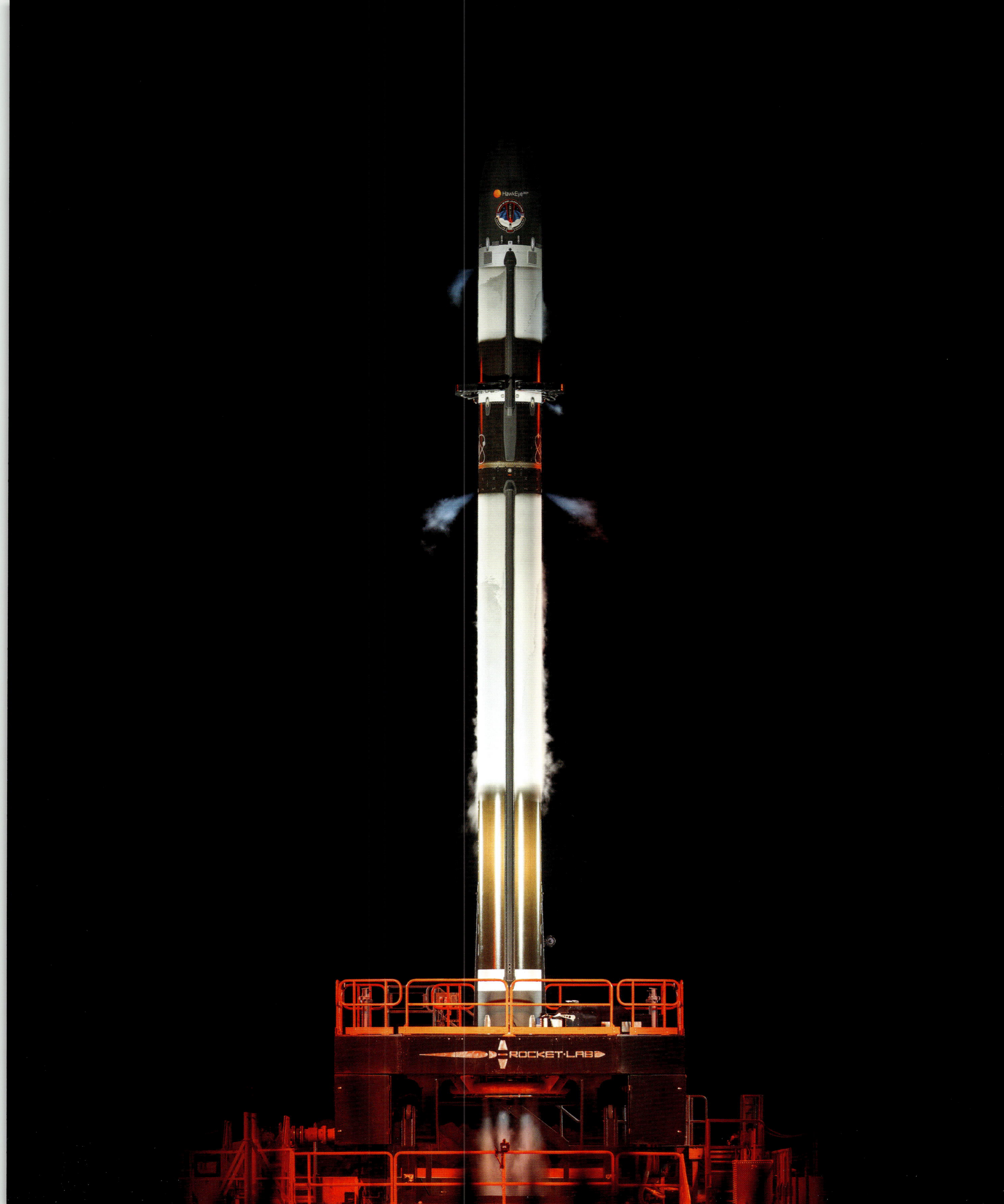
HawkEye
ROCKET·LAB

Oh, the Humanity!

2018 was a pivotal year for Rocket Lab, marking its transition from testing to commercial operations. Peter Beck and his team would take eight months to follow up "It's a Test" with another launch—a rapid turnaround by industry standards following a first launch.

The "Still Testing" launch took place in January 2018, after a series of weather, orbital traffic, and operational delays pushed the launch into the new year. The mission title cemented the quirky naming convention Rocket Lab would use for all subsequent launches.

"We have to log the missions with Space Command in the US, and they track the rocket as it's flying into orbit," said Peter. "They require a name so they can put it on their screens, so in typical dry fashion, we called the first one 'It's a Test.'"

It changed how rocket missions were named throughout the industry, and would help build interest in the launches among a rapidly growing section of the public interested in tuning in to watch them livestreamed on the internet.

"Still Testing" signaled to the world that Rocket Lab was keen to gather more data to perfect its technology and launch systems. But there were also previously unannounced commercial payloads onboard consisting of a Dove Earth-imaging satellite for Planet Labs, and two Lemur-2 CubeSats for Spire Global, intended for weather- and ship-tracking.

Not wanting to get over its skis, as Peter would put it, Rocket Lab was also keeping a couple of other secrets about its second launch.

Curie and Electron's Kick Stage

"Still Testing" was the first time the Electron Kick Stage was used. A versatile third stage of the launch vehicle, it was designed to provide precise orbital deployment and enhanced mission flexibility.

The Kick Stage is powered by another Rocket Lab 3D-printed engine, Curie, named after the physicist and chemist Marie Curie, who conducted pioneering research on radioactivity.

During the second mission, the Kick Stage deployed the satellites into circular orbit. Around forty minutes later, Mission Control activated Curie, successfully firing up the engine.

"The Kick Stage even has its own precision gas thruster control system, avionics package, power source, and communications system. The small stage uses its propulsion system to de-orbit after deploying its payload, reducing the amount of space junk left behind by launches," *Popular Mechanics* reported following the launch.

The Kick Stage was left in orbit, but for subsequent launches the stage conducted an orbit-lowering burn to speed up its deorbit after releasing the payloads. The Kick Stage became an integral part of Rocket Lab's launch services, allowing satellites to be delivered to unique and precise orbital destinations. Curie can also perform multiple burns for more complex missions.

Rocket Lab was officially in the commercial satellite launch business. But it had another card up its sleeve with "Still Testing." An additional secret payload was onboard—Humanity Star.

A 3.2-foot-wide (1 m) geodesic sphere covered in highly reflective panels, the silver disco ball was designed to be visible from Earth as a bright, moving point of light. It was conceived by Peter to make people all over the world "look up."

Astronauts looking down at Earth from space describe the "overview effect" as the sense of Earth's fragility and loneliness in the solar system when viewed from the Moon. It often inspires them to recognize the importance of protecting our planet and each other. Peter believed most Earth-bound humans could experience the same sense when they peered up at the night sky.

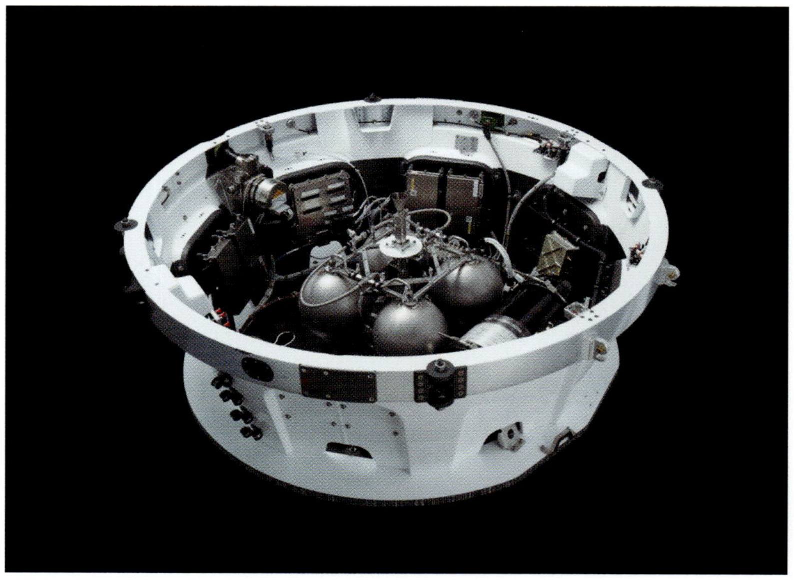

Above: The Kick Stage on orbit.

Right: Rocket Lab's Kick Stage.

This page: The Kick Stage's Curie propulsion system.

"... when you look up, you realize how totally insignificant you are in the universe. And all of a sudden you start asking the crazy questions like, 'Where are we from?'"

— Sir Peter Beck

The Humanity Star served as a reason to encourage people to go outside and look up.

"It's a little depressing, but when you look up, you realize how totally insignificant you are in the universe. And all of a sudden you start asking the crazy questions like, 'Where are we from?'" he said.

A Sputnik Moment

In October 1957, the Soviet Union launched the world's first satellite, Sputnik, into orbit, sending shockwaves through NASA and triggering the space race. For twenty-two days, before Sputnik's batteries ran out and it deorbited and burned up in Earth's atmosphere, people around the world could literally hear its pulse.

Shortwave radios picked up the distinctive and now iconic "beep, beep" of Sputnik as it passed overhead.

"I wanted to do my version of Sputnik," said Peter. But it would be a visible, rather than an audible, statement. Humanity Star was successfully deployed. As it separated from the Kick Stage, it was deliberately spun on two axes, sending it into a continuous tumble, all the better to flicker in the night sky.

"It would be this kind of disco ball, sparkling, and glistening. Even if you never saw it, if you went outside and looked up, my job was done," he reflected.

Humanity Star served its intended purpose, perhaps too well. It was observed from all over the world during the three months it was in orbit. But it triggered intense debate about the light pollution and space junk that was likely to accompany the ramping-up of commercial satellite launches, which was soon to reach a new threshold, with SpaceX gearing up to launch a constellation of broadband satellites.

To Peter's annoyance, the criticism gained particular momentum in his home country.

"There was one particular New Zealander astronomer who created a really large stink about it polluting the night sky," he remembered. "There was a bunch of negative media generated from that one ice crystal in the pond. Ironically, it came from New Zealand."

Regardless, he doesn't have any regrets about Humanity Star. While attending a conference a few weeks after the launch, he got up before dawn with a group of other delegates to watch Humanity Star pass overhead.

"It was such a cool thing, to share that moment," he said. "We even heard about one lady who had Humanity Star tattooed on her left ass-cheek. That's dedication, that's how powerful it was to some people."

Three months later, Elon Musk would deposit a Tesla Roadster into orbit around the Sun, where it will remain for tens of thousands of years. "There didn't appear to be nearly as much stink about that," Peter complained.

It's Business Time

The year would end with two more launches, including "It's Business Time," Rocket Lab's first fully commercial launch, and a tip of the hat to the quirky New Zealand pop band Flight of the Conchords, who had a global hit with the song of the same name.

Launch number three again used the Curie engine-powered Kick Stage, and carried satellites for Spire Global and Planet Labs. Originally scheduled for April, the launch was delayed several times, largely due to a pesky motor controller issue.

Just five weeks after "It's Business Time," the Electron rocket would take flight again, marking a major milestone for Peter and the company—it was Rocket Lab's first mission to launch satellites for NASA. "This One's For Pickering" honored the New Zealand-born former Jet Propulsion Laboratory Director, Bill, who died in 2004, aged ninety-three.

NASA has been launching satellites as part of its Educational Launch of Nanosatellites (ELaNa) science, technology, engineering, and mathematics (STEM) education program since 2011. One of the CubeSats was Repair Satellite-Prototype (RSat-P), which was built by the US Naval Academy in Annapolis, Maryland.

It was designed to perform minor repairs on a larger satellite, addressing the disastrous scenario where an orbiting satellite stopped working, often due to a solar panel malfunctioning.

Alongside the ten small satellites released were two identical CubeSail CubeSats, an experimental spacecraft propulsion system that, when in orbit, was programmed to deploy a 215-square-foot (20 sqm) solar sail ribbon between them. The radiation pressure from Sunlight can be enough to propel a satellite in zero gravity, with the sail designed to catch the Sun's energy along its surface.

With such a short period between flights three and four, Rocket Lab demonstrated it could achieve a rapid launch cadence as it

Previous page: The Humanity Star.

Above: The Kick Stage in space with its payload for Electron's ninth launch, "As The Crow Flies," October 2019.

looked to increase the frequency of missions. The year ended with a flurry of activity, not only on the launchpad at Mahia, but in Rocket Lab's boardroom, as Peter and senior executives closed the company's Series E financing round, raising $140 million. That brought the total funds raised to $288 million, and valued Rocket Lab at over $1 billion.

Existing backers, Future Fund (an Australian sovereign wealth fund), Greenspring Associates, Khosla Ventures, Bessemer Venture Partners, DCVC (Data Collective), Promus Ventures, and K1W1 all injected new investment into Rocket Lab, and New Zealand's Accident Compensation Corporation (ACC) joined the round.

"This funding enables the continued aggressive scale-up of Electron production to support our targeted weekly flight rate. It will also see us build additional launchpads and begin work on three major new R&D programs," Peter said in a press statement.

The teams in Huntington Beach, California and Auckland began the complicated task of ramping up production. By the end of 2018, Rocket Lab had launched a total of twenty-four satellites into orbit, and signed agreements for multiple future launches, including a contract with launch aggregator Spaceflight for three launches.

LC-1 at Mahia had proven its value, but plans were afoot for a second Rocket Lab launchpad, half a world away, and on the doorstep of some of Rocket Lab's key target customers.

Mission Patches

The art of space mission patches dates back decades to the earliest human spaceflight missions, and are a time-honored tradition that has been adopted by Rocket Lab since its first launches to space with Electron. Each mission patch includes design elements that represent the customer, spacecraft, or mission objective of that Electron launch, along with the launch's mission name.

01

02

03

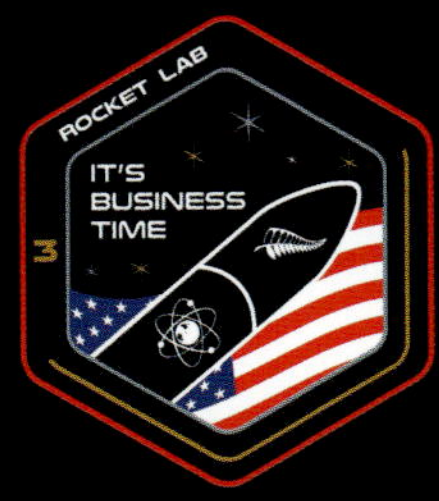

04

05

06

07

08

09

10

11

12

13

14

15

16

17

18

19

20

21

22

23

24

25

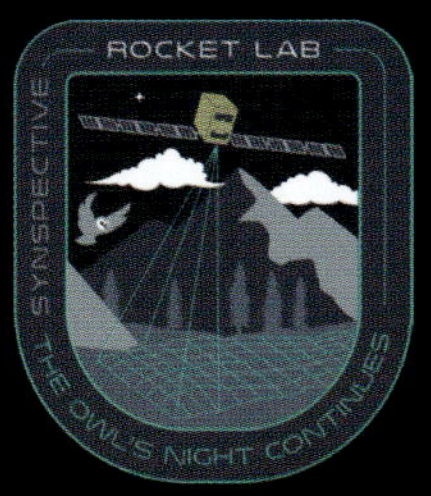

26

27

28

29

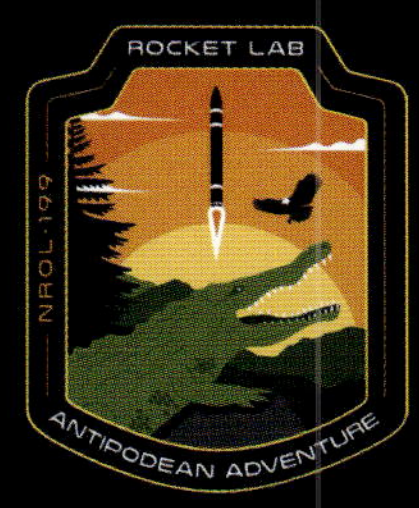

30

31

32

33

34

35

36

37

38

39

40

41

42

43

44

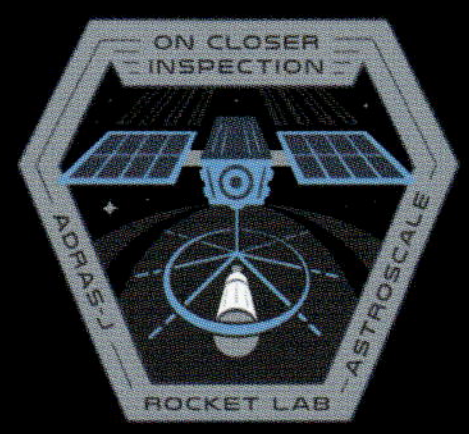

45

46

47

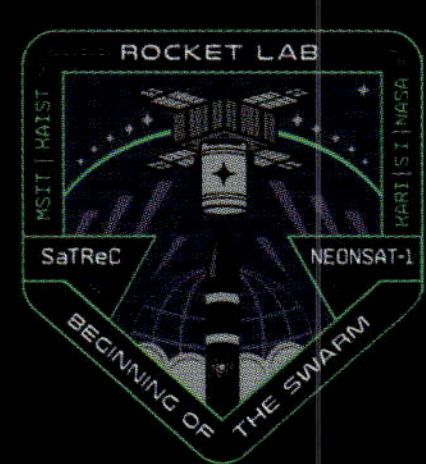

48

49

50

ELECTRON REACHED ITS FIFTIETH LAUNCH FASTER THAN ANY OTHER COMMERCIALLY DEVELOPED ROCKET IN HISTORY AS DEPICTED ON THE GRAPH BELOW.

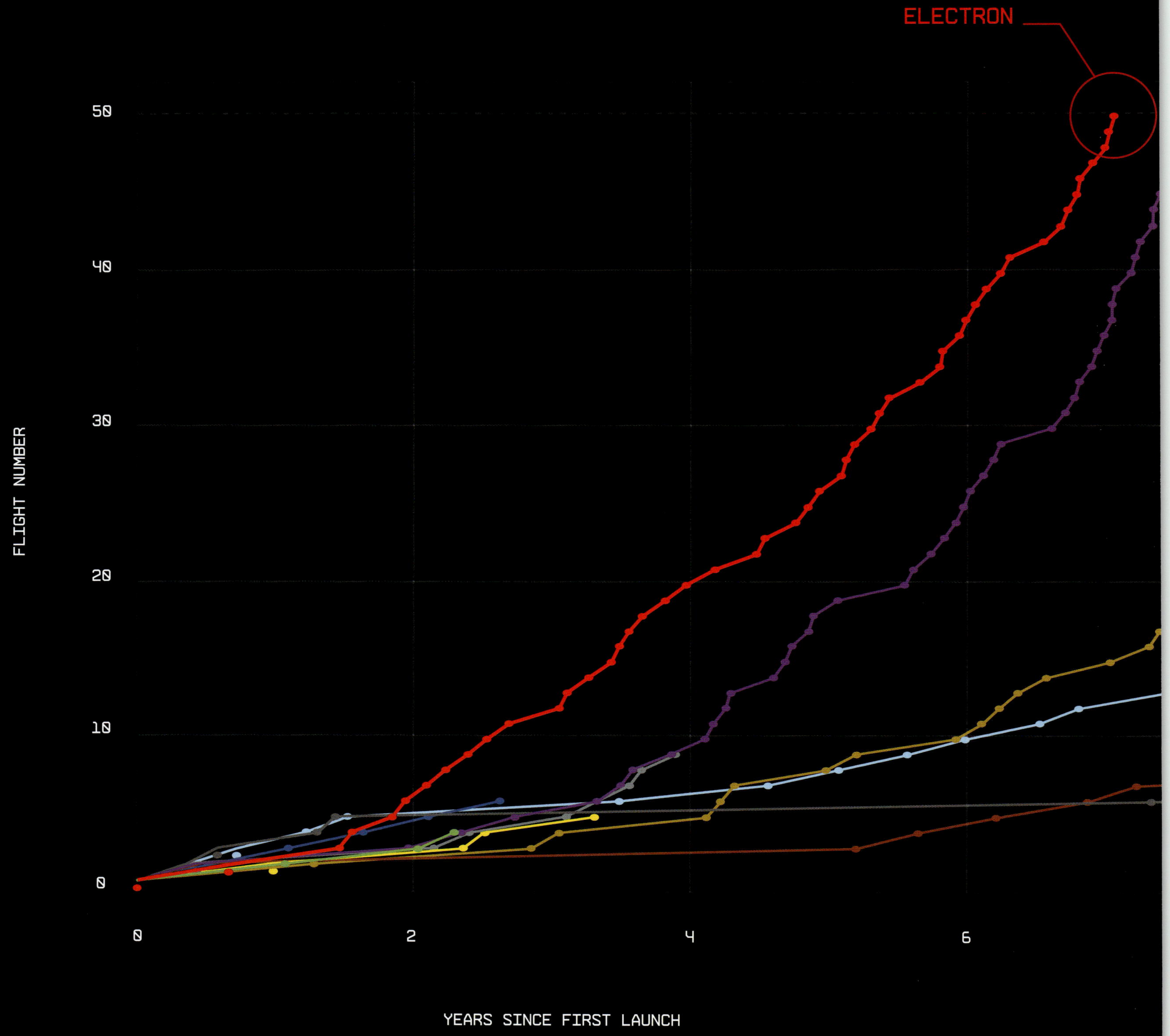

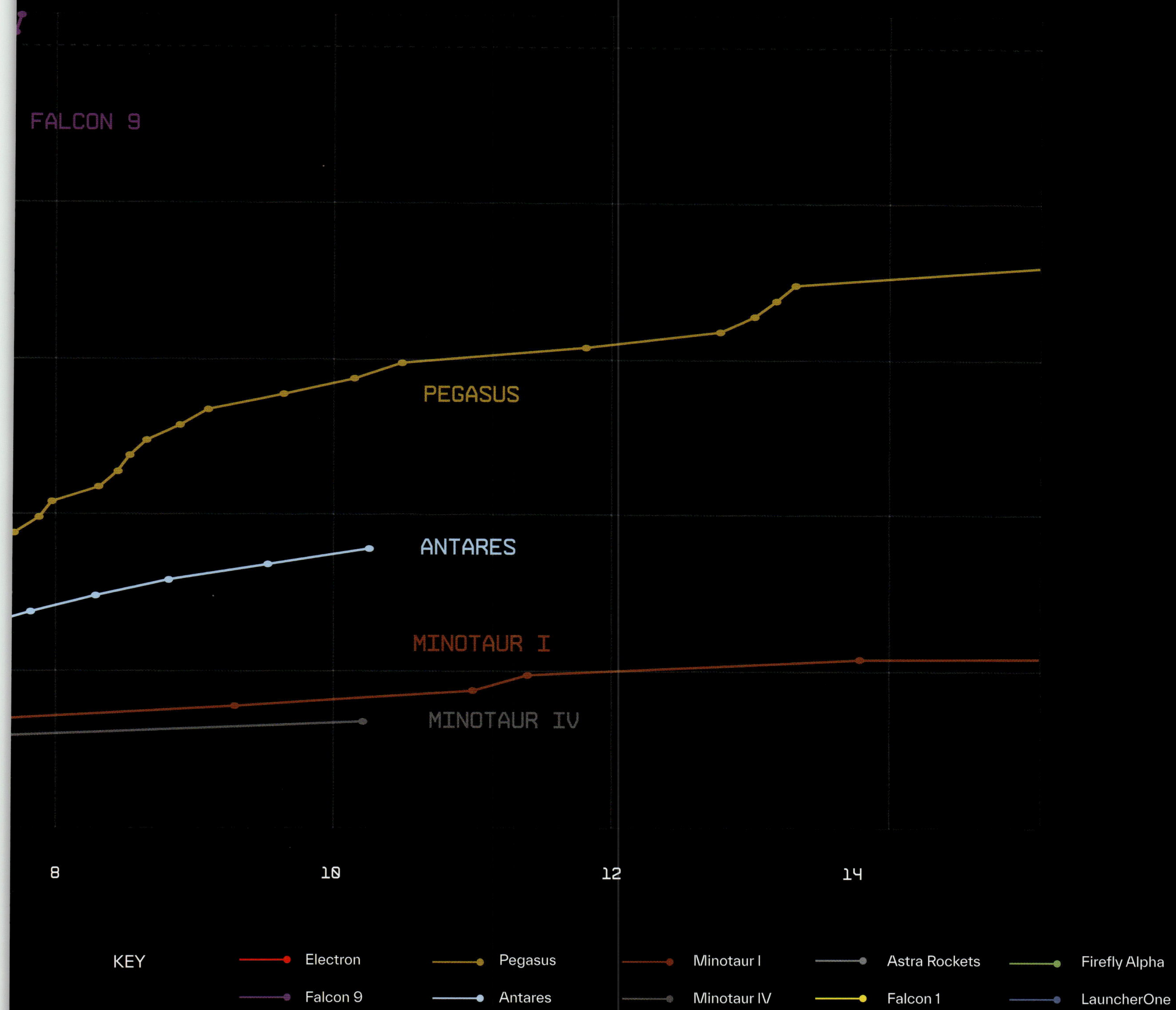
FALCON 9
PEGASUS
ANTARES
MINOTAUR I
MINOTAUR IV
8
10
12
14
KEY
Electron
Falcon 9
Pegasus
Antares
Minotaur I
Minotaur IV
Astra Rockets
Falcon 1
Firefly Alpha
LauncherOne

LAUNCH COMPLEX:

2

Rocket Lab's second launch site, Launch Complex 2, represents a new responsive launch capability for the United States on home soil, and is located on Wallops Island, Virginia.

Tailored specifically for US government small satellite missions, Launch Complex 2 can support up to twelve missions per year.

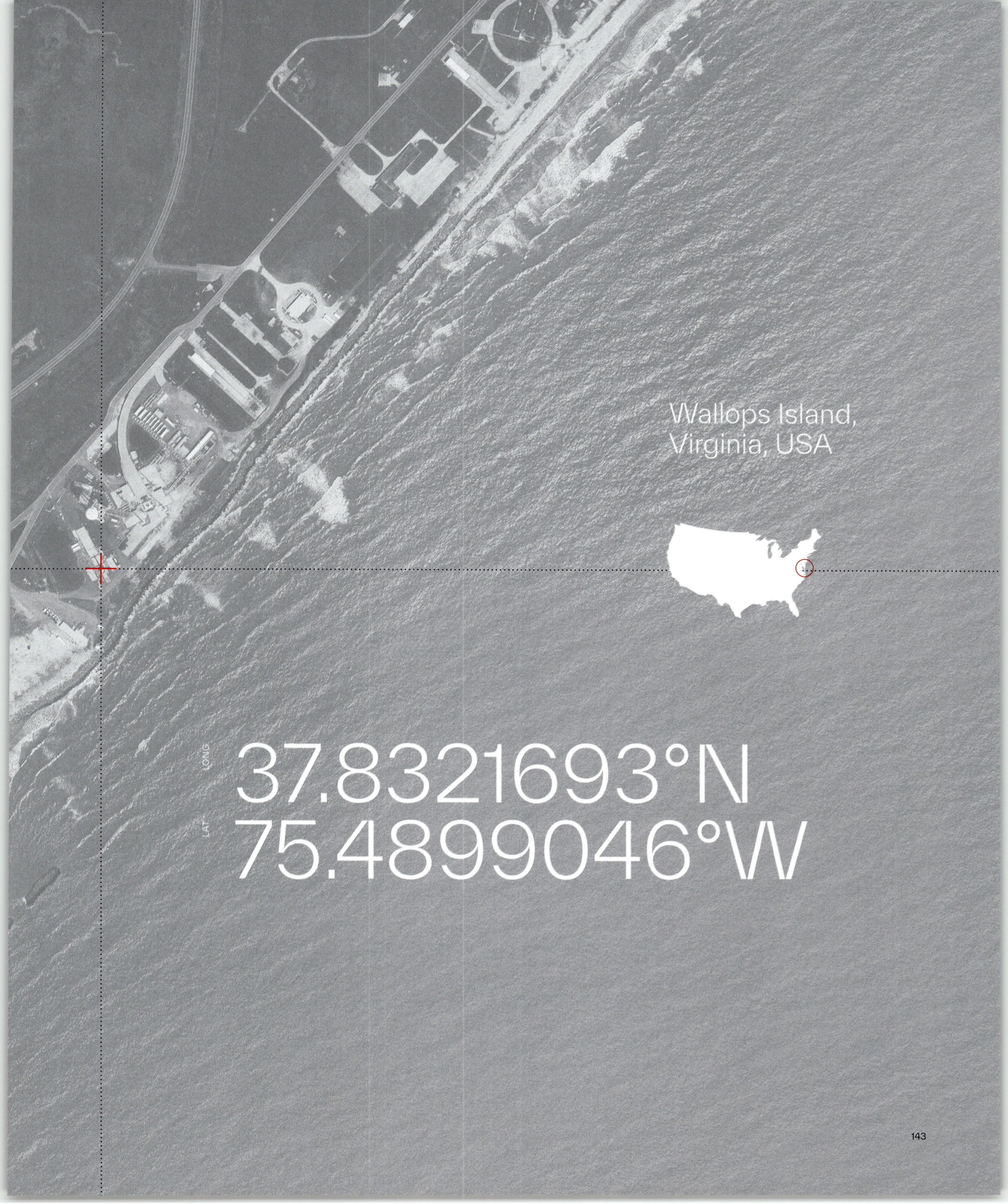

Wallops Island,
Virginia, USA

37.8321693°N
75.4899046°W

ELECTRON

Launching From America—LC-2 at Wallops Island

The Mid-Atlantic Regional Spaceport at Wallops Island, Virginia, straddles a sprawling barrier island chain along the Atlantic coast. It's a diverse landscape with beautiful sandy beaches on the seaward side and a complex tidal marsh system on the landward side that covers over 3,330 acres (1,335 hectares).

The facility is one of the few in the US authorized to conduct Low Earth Orbit (LEO) launches, and has hosted many resupply missions to the International Space Station. Owned by the Commonwealth of Virginia, the complex's anchor tenant is NASA. Its history as a launch site stretches back to 1995 when Conestoga 1, the world's first privately funded commercial rocket, first launched in 1982, took its final flight. The rocket exploded forty-six seconds into its flight and the Conestoga space program died with it.

Around forty commercial launches have taken place at Wallops Island. Defense contractor Northrop Grumman has been responsible for the lion's share of them, with its medium-lift Minotaur and Antares rockets undertaking missions for NASA and the National Reconnaissance Office.

With its legacy hosting commercial operations, it was an obvious choice for Rocket Lab to develop a second launch complex. The reason for doing so came down to satisfying the needs of its major customer—the US government.

"They asked us to do it, and even paid for us to develop the site," Peter Beck explained. "We needed a place where we could launch national

> "We turned up to a NASA facility, a government facility, and literally hoisted a pirate flag. Everybody rose to the challenge and got where they needed to get to."
>
> — Sir Peter Beck

Left and previous page: An Electron rocket stands vertical on the pad at Rocket Lab Launch Complex 2.

Opposite and overleaf: From a distance: Rocket Lab Launch Complex 2.

security payloads from US soil."

Launch Complex 2 (LC-2) would allow for rapid call-ups—urgent missions that could be arranged in a matter of weeks, rather than months or years. Rocket Lab did due diligence on several US launch sites, including Florida's Cape Canaveral, Alaska's Pacific Spaceport Complex, and California's Vandenberg Space Force Base.

Wallops Island had the edge with the ability to accommodate frequent launches, existing infrastructure in the form of NASA's Wallops Flight Facility and the Mid-Atlantic Regional Spaceport, and the potential for a rapid construction timeline. The latter was particularly important with Rocket Lab targeting a US maiden launch for the third quarter of 2019.

There were geographic advantages to Wallops Island, with its position on the eastern seaboard making it ideal for supporting various orbital inclinations and potential return-to-launch-site capabilities for future reusable rockets.

But it was also close to important government agencies and military and aerospace contracting companies that Rocket Lab did business with. The real clincher, however, was the high launch frequency the site could support—at least twelve missions per year. That kind of cadence wasn't possible for a newcomer at other US launch sites, due to congestion with other rockets.

Building a New Pad

Construction on LC-2 began in February 2019 and moved at an impressive pace. The LC-1 build team was reassembled and deployed to Virginia, alongside a handful of local hires, to build the new launch site and complete its construction within ten months.

Shaun D'Mello, who had scoured New Zealand five years earlier looking for a suitable site for LC-1, and was now Vice President—Launch at Rocket Lab, played a pivotal role in overseeing development at Wallops Island.

The physical infrastructure was in place faster than an average four-bedroom home would take to build back in New Zealand. However, the Covid-19 disruption interfered with further development of the site, and obtaining the necessary permits and licenses to operate the launch site pushed out the timeline. Seeing as NASA controlled the Wallops Flight Facility, it required Rocket Lab to use its autonomous flight termination system (AFTS) for range safety. An AFTS differs from the flight termination system utilized in the early launches from LC-1, which relied on a human-in-the-loop approach. In that earlier system, a person was required to monitor the launch and manually activate the termination switch if the rocket deviated from its intended course or behaved abnormally. With an AFTS, the system is entirely automated within software on the rocket, enabling it to make decisions faster and with more precision.

"There's a very long certification process that, quite frankly, we probably underestimated how long it would take," Peter told *Space News* in August 2020. "There were a lot of struggles there in the early days," he later reflected.

The pace at which the Rocket Lab team liked to move wasn't always realistic at a sensitive site dominated by NASA and multiple other parties.

Jane McNicoll, Manager—Mission Management at Rocket Lab, explained that launching from Wallops Island satisfied the "tight requirements" for certain US government satellite launches, but was a more complex process.

Jane McNicoll, Manager—Mission Management, Rocket Lab.

“That was a very successful launch, but it was a challenge to get there on launch day, because Virginia in the middle of winter is an environment that we’re not familiar with here in New Zealand.”

— Michael Pearson, Director—Test and Launch Operations, Rocket Lab

Right and overleaf: Successful lift-off of an Electron rocket from Rocket Lab Launch Complex 2.

“LC-1 at Mahia feels like you’re in your backyard doing whatever you want. At LC-2 there’s so many people to manage, and everybody has their own piece of the pie. That’s a lot to balance,” said Jane, who began her career at SpaceX and worked there as a Systems Engineer on the fledgling Starlink project.

The first launch at LC-2 would not happen until January 2023. Michael Pearson, who oversees the integrated operations team tasked with performing test and launch operations for the Electron rocket, counts that first launch from Wallops Island, dubbed “Virginia is for Launch Lovers,” as his favorite.

“That was a very successful launch, but it was a challenge to get there on launch day, because Virginia in the middle of winter is an environment that we’re not familiar with here in New Zealand,” he said. “The first time is always scary. We weren’t entirely sure of how the vehicle was going to behave in that sort of weather, and all of the external factors.”

It was also his first time working closely with NASA. “Cumulatively, I probably spent about six months over there working on that first mission to get that one off the pad.”

A Home for Neutron

It was a proud moment for everyone concerned to watch Electron blast off at 6 p.m. EST on January 24, 2023. Three satellites were deployed to an orbit of 341 miles (550 km) for Virginia-based HawkEye 360, a geospatial analytics company that collects and analyzes radio frequency signal locations using a constellation of satellites built up over the course of numerous launches.

It represented an innovative use of technology that spoke to Peter’s desire to give satellite companies the ability to better understand what was going on back on Earth. HawkEye 360’s analysis of radio frequency signals can pinpoint the location of ships even when their Automatic Identification System (AIS) has been turned off. It can also quickly locate signals from emergency beacons, aiding in search and rescue efforts.

The launch brought the total number of satellites Rocket Lab had launched from its three launchpads to 155.

“After our busiest launch year yet in 2022 with nine successful missions, what better way to kick off the new year than by launching Virginia-built spacecraft from a Virginia launch site, enabled by our rapidly growing Virginia-based team,” Peter said in a press statement.

The confidence expressed in the press release masked the convoluted process involved in getting LC-2 ready to host launches.

“We turned up to a NASA facility, a government facility, and literally hoisted a pirate flag,” Peter summed up. “Everybody rose to the challenge and got where they needed to get to.”

While LC-2 was initially developed for Electron launches, its strategic location has proven valuable for Rocket Lab’s ambitious plans with its larger Neutron rocket.

In February 2022, Rocket Lab announced that Wallops Island would also be the site for Neutron’s launchpad. The dedicated Neutron launch site, known as Launch Complex 3 (LC-3), has been built adjacent to LC-2. LC-3 is designed to support Neutron’s reusability features, including return-to-launch-site landings and refurbishment for rapid reuse. Rocket Lab tests the Archimedes engines that power Neutron and other components across various facilities in the US.

It is unlikely that Neutron rockets will ever be launched from the original Mahia Peninsula site. “It’s just not feasible to do,” said Peter. “You could take all of the liquid oxygen in the country and half fill the tank once. Nor is it the right place to have a medium-lift launch vehicle, when your customers are in North America.”

Instead, the three launch complexes give Rocket Lab ultimate flexibility as Neutron’s debut nears. From small satellites deployed by Electron to larger payloads carried by Neutron, Rocket Lab now can launch the majority of addressable satellites in the space industry.

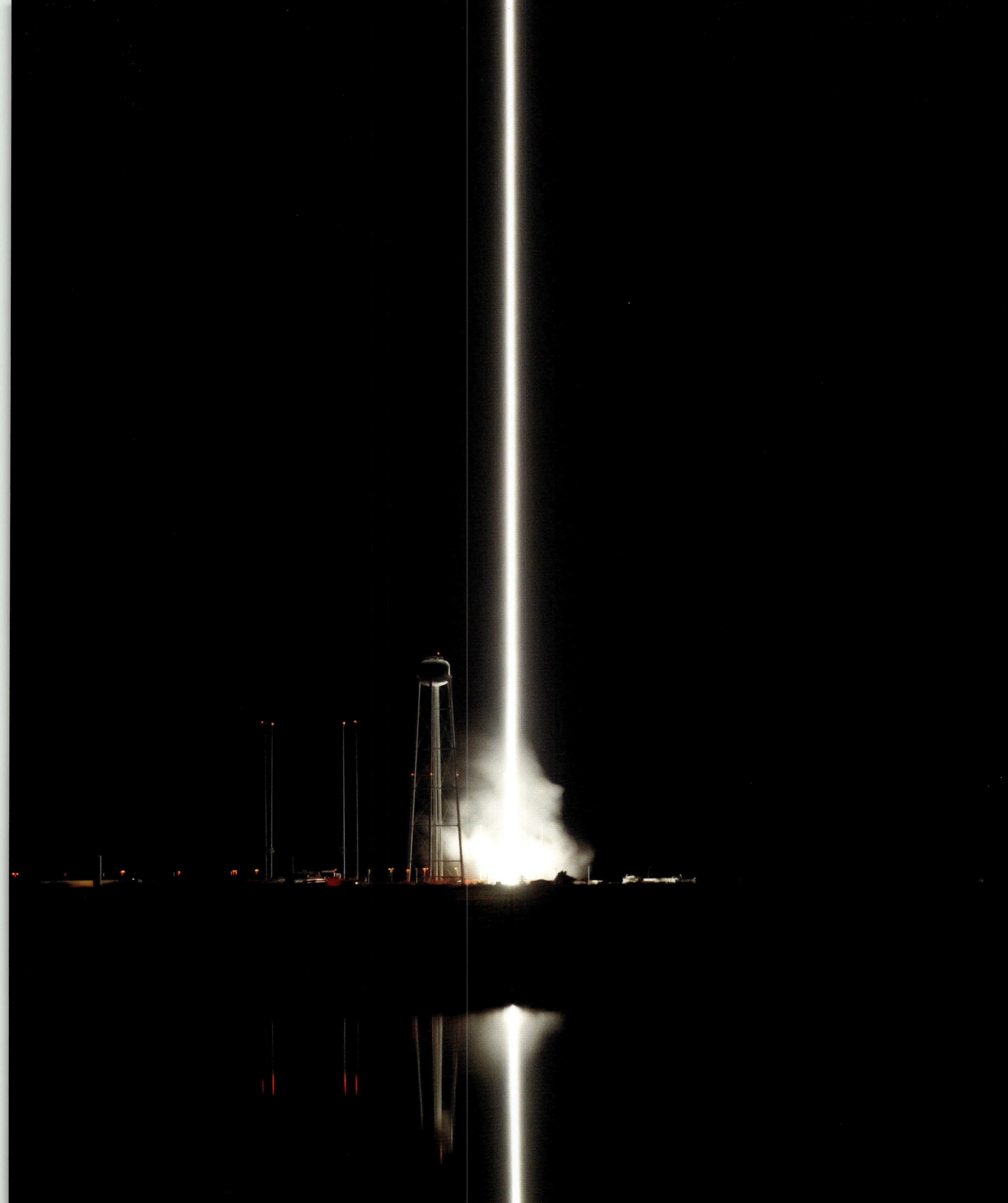

NASA

Rocket Economics— To Use and Use Again

Right: A mid-air helicopter capture test operation with an Electron test article.

From recovering rocket stages that have splashed down in the ocean, to landing them upright on a launchpad or barge, a massive engineering effort has been devoted to making rocket components reusable.

> "The rocket had to launch precisely at the right time. It had to reenter the Earth's atmosphere at exactly the right place. Then the helicopter had to be within the safety zone, rendezvous with the rocket and match its descent."
>
> — Sir Peter Beck

Rocket Lab has embraced rocket reusability. The rationale for it comes down to lowering costs and the ability to increase launch frequency. In the small rocket business, where margins are typically thinner than for larger launch vehicles, reusing engines and a rocket booster can significantly lower the overall cost of a launch.

Refurbishing rather than manufacturing rocket components also speeds up the production timeframe, allowing ready-to-use rockets to be delivered to the launchpad at a higher cadence.

In 2018, Rocket Lab announced plans to develop a reusable first-stage booster, as it looked to increase the frequency of Electron launches. Two methods of salvaging the Electron booster were planned—ocean recovery and mid-air capture. The first involved returning the booster to Earth under a parachute and splashing it down in the ocean. A specialized recovery boat would then retrieve the rocket stage and take it back to land.

The second, more difficult method was mid-air capture. Rocket Lab planned to use a helicopter to hook the rocket stage as it descended under its parachute, towing it to land where it could be settled upright back on the launchpad.

Peter Beck knew early on in Rocket Lab's history that without reusable rockets, the company would struggle to compete, particularly as it pursued its plan to offer mid-lift rocket capability.

"If you're not building a reusable rocket, it's a dead-end product," he said. "There's a caveat for small launch vehicles," he added. "There are certain elements of the recovery that are fixed costs, such as the recovery boat."

For Electron, Peter ultimately moved on to a more precise method—the mid-air catch. There was a historical precedent for it. Between 1959 and 1972, the Central Intelligence Agency (CIA) and US Air Force operated the CORONA program to retrieve film from spy satellites. The secret program was deemed essential when a U-2 spy plane flying over the USSR to photograph suspected missile sites was shot down by the Soviets.

The spy plane's CIA pilot, Francis Gary Powers, was captured near Sverdlovsk in west-central Russia. President Dwight D. Eisenhower stopped all U-2 flights soon after, forcing the CIA to come up with a new way to gather intelligence about Russian weapons capabilities.

Launching a satellite into orbit to take photos of Russia was relatively straightforward. Retrieving the film was another story entirely. In the case of the CORONA missions, the satellite reentering Earth's atmosphere would deploy a capsule, nicknamed "film bucket." It would deploy a parachute and descend to around 15,000 feet (4,572 m) at which point a plane towing a big claw would hook the capsule's parachute and either winch it into the plane or drop it in the ocean for easy recovery.

CORONA was a great success. In a single mission, it produced more photographic coverage of the Soviet Union than all previous U-2 missions combined. But no one had ever caught a rocket mid-air, which would require use of a helicopter rather than an aeroplane.

"Mid-air capture wasn't crazy; it had been demonstrated. The fundamentals seemed to be there," said an undaunted Peter.

Success would ultimately come down to perfect timing. "The helicopter only has a certain amount of loiter time, because it's [248 miles] 400 km out over the ocean," Peter, a helicopter pilot himself, figured.

"The rocket had to launch precisely at the right time. It had to reenter the Earth's atmosphere at exactly the right place. Then the helicopter had to be within the safety zone, rendezvous with the rocket and match its descent."

IXW

Supersonic Ballet

As 2021 progressed, Rocket Lab developed its approach to mid-air capture, coming up with a method of hooking the rocket stage via its parachute line, and figuring out a process for Mission Control to coordinate the rendezvous of the helicopter with the descending rocket stage. "We were super lucky to find arguably the best long-line pilot in the entire world. He was able to master that pretty quickly," said Peter.

In April 2020, as Covid-19's spread compelled the New Zealand government to shut its borders, a major test of the mid-air recovery method took place over the Pacific Ocean. A helicopter carried a dummy rocket booster, while another deployed a cable and hook to attempt to capture it.

The helicopter with the dummy load released it at around 8,000 feet (2,438 m), with a parachute deploying to slow its descent. Then the second helicopter swooped in and at 5,000 feet (1,524 m) hooked the parachute cable on the first attempt. It then landed the dummy rocket on the mainland.

Peter described the drop test as a "logistical nightmare" to CNBC. "We had two large vessels, a recovery boat, and three helicopters. So it was quite a flotilla of vessels that were going out into the middle of the ocean to do this," he said.

But it proved mid-air capture was possible. Following more testing, everything was in place to attempt a real Electron stage mid-air recovery two years later. It would happen at 10.49 a.m. NZST on May 3, 2022, with the "There and Back Again" mission, another New Zealand-themed name for the logbook, with the country having been the filming location for Sir Peter Jackon's epic *The Hobbit* and *Lord of the Rings* trilogies two decades earlier.

Electron's twenty-sixth mission went off without a hitch, deploying thirty-four satellites into orbit, including SpaceBEE Internet of Things (IoT) satellites for Swarm Technologies, and PocketQube research satellites for Alba Orbital.

But the atmosphere at Mission Control remained tense as the Electron booster drifted down under its parachute. At 6,500 feet (1,981 m), Rocket Lab's red, white, and blue Sikorsky S-92 helicopter with a crew of two pilots and a spotter was waiting with its long-line and hook deployed.

Another helicopter hovering at a distance and cameras on the Sikorsky captured the moment the hook snagged the parachute line and began towing the rocket. A huge roar of celebration went up at Mission Control, where staff had packed into the viewing area to watch the historic event. The delicate maneuver worked, but it would be a brief union.

After successfully catching the rocket stage, the pilot calculated that his cargo had slightly different load characteristics than those experienced during testing. In the interest of safety, he jettisoned the rocket, which splashed down and was quickly recovered by a boat that was standing by.

Nevertheless, Peter was exhilarated. Another world first had been achieved. "A tremendous number of factors have to align, and many systems have to work together flawlessly, so I am incredibly proud of the stellar efforts of our recovery team and all of our engineers who made this mission and our first catch a success," he said in a media statement.

"If you're not building a reusable rocket, it's a dead-end product."

— Sir Peter Beck

Above and right: Operations to catch an Electron test article mid-air with a helicopter proved successful and paved the way for future mid-air recovery attempts during Electron missions.

Opposite: Rocket Lab's customized Sikorsky S-92, a large twin-engine helicopter typically used in offshore oil and gas transport and search and rescue operations.

Splashdown Isn't So Bad after All

The rocket stage was returned to Rocket Lab's production facility in Auckland for extensive testing and analysis. Another mid-air catch was attempted in November 2022 at the conclusion of the "Catch Me If You Can" mission, which successfully deployed a 176-pound (80 kg) Swedish research satellite into space.

Again, the Sikorsky was poised to move in with its hook. However, a loss of telemetry data from the descending rocket stage led to the effort being abandoned, as it wasn't safe enough for the helicopter to remain in the recovery zone without the crew knowing exactly where the package was. Again, the Electron stage splashed down and was recovered.

It was the analysis and testing of that stage that led to a pivot in Rocket Lab's approach to reusability. Despite being dunked in saltwater, the booster was in good shape. Why not just make splashdown recovery Rocket Lab's go-to method?

"Every time we brought a stage back to the lab, it was kind of, 'Actually this isn't that bad, do we still need the helicopter to do this?'" Peter said. "It got to the point where we were improving the stage to the extent that we could just seal it up to a greater degree to waterproof it."

There would be no more mid-air catches. But since 2022, Rocket Lab has recovered dozens of Electron first stages from the water. In a significant milestone, Rocket Lab successfully reflew a previously used Rutherford engine on its fortieth Electron mission in August 2023, demonstrating progress in engine refurbishment and reuse.

Around 2024, Rocket Lab put Electron's reusability program on ice to focus on scaling rocket production to meet growing customer demand for dedicated small launch. But the company's progress on reusable components for Electron has generated a wealth of experience and data that's invaluable as the company aims to put reusability at the heart of its Neutron program.

"I pity anybody who goes into a reusable program stone cold," said Peter. "There's a lot of stuff you can design, a lot of stuff you can analyze. But there's a lot of stuff where you just have to fly. Electron gave us the ability to do flight testing. So when it came to Neutron, it was not scary."

Unlike Electron, Neutron is designed for propulsive landing of the first-stage booster, complete with its nine Archimedes engines, back at the launch site, similar to SpaceX's Falcon 9. Neutron also features a unique "Hungry Hippo" fairing design, a nose cone that remains attached to the first stage, eliminating the need for separate fairing recovery. That takes reusability to another level.

"We wanted to basically retrieve everything that was suborbital in one piece—so everything that doesn't go to orbital velocities. That way we wouldn't have multiple marine operations that simply drive up cost," explained Shaun D'Mello, Vice President—Neutron.

Typically, the fairings separate to clear the way for the payload to be deployed and are either destroyed in Earth's atmosphere or splash down into the ocean. The nose cone protects the payload from the extreme pressure and heat encountered as the rocket blasts through the atmosphere. The fairings are specialized and expensive rocket parts. If a fairing fails to open properly, it can lead to the mission failing.

Shaun and the Neutron team came up with a design that was first known as the Crocodile Fairing, and then the Hungry Hippo. It involves the fairing being attached to the rocket's first stage. Instead of falling away when they open, the fairings remain attached, opening up like a yawning set of jaws to release the second stage and the payload.

"You could deploy the second stage and then close the fairings and bring the whole rocket back down," said Shaun. "It's all in the name of not having to go and fish for more fairing parts out in the ocean, but it's also about not having to reassemble major parts of the rocket. Just like a big cargo plane—you don't go and reattach the doors after every mission."

Like the Hungry Hippo, Rocket Lab has incorporated reusability into other aspects of Neutron's design. The rocket will use a specially formulated carbon composite material designed to withstand multiple launches and reentries. The new Archimedes engine for Neutron features an oxidizer-rich closed cycle for improved engine life.

Reusability is now part of the business model of many rocket companies, improving the economics of getting to orbit and speeding up launch cycles. It's also tackling the problem of space debris, a factor facing growing scrutiny as the number and frequency of rocket launches continues to increase.

Below left and right: An Electron first stage awaits recovery from the ocean after successfully returning from space during the "Four of a Kind" and "There and Back Again" missions in January 2024 and May 2022 respectively.

Above: Electron's first stage is hoisted from the ocean.

Below: Electron onboard Rocket Lab's repurposed recovery vessel.

"Every time we brought a stage back to the lab, it was kind of, 'Actually this isn't that bad, do we still need the helicopter to do this?'"

— Sir Peter Beck

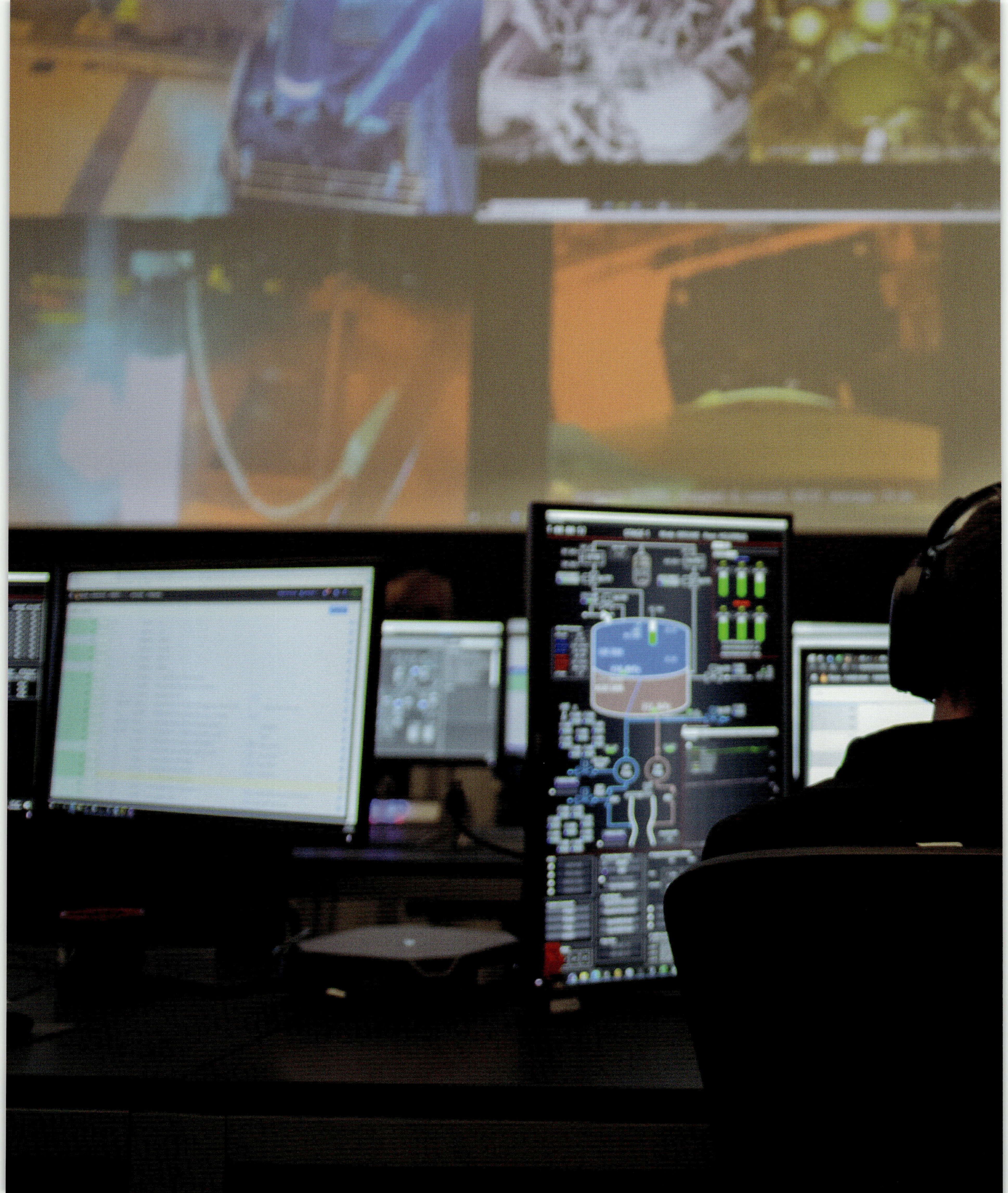

The Precise Art of Launching a Rocket

A fresh-faced Peter Beck personally counted down the Ātea-1 launch on Great Mercury Island back in 2009, and assumed the role for numerous subsequent launches.

But these days you'll rarely find him in Mission Control on launch day. "I hate launch day," he admitted. "There's always a tremendous amount of responsibility that goes along with that. All of these missions, they're not for fun, they're really important, and there's a lot riding on them."

He will usually watch on from his desk, leaving the complicated job of launching a rocket to a sizable team of experienced experts with nerves of steel, and using procedures refined over the course of sixty-eight Electron missions (as at time of writing).

But the groundwork for success on launch day is laid months or years earlier, and involves extensive collaboration between Rocket Lab and the customer whose satellite payload will be carried into orbit.

The person responsible for juggling customers' demands and the needs of Rocket Lab's engineering team is Jane McNicoll, Manager—Mission Management.

"With a lot of other launch vehicles, you get what you get, you're just another person on the bus," she said. "When you come to us with your satellite, we're open to making crazy changes and offering really unique solutions. It might be altering the fairings to fit the satellite into the rocket. We will tailor our solution to meet the need."

The process of preparing the rocket and payload for an Electron launch used to span a nine- to twelve-month period. Recently, however, Rocket Lab has turned around missions from signature to launch in as little as two months. Jane pays particular attention to explaining what the customer's plans are for the payload once it's in orbit.

"It drives home that, at the end of the day, we're all here to serve this purpose. The science mission of what we're trying to do here is special and meaningful every single time."

Julia Wembacher, Team Lead—Integrated Operations Team X, Rocket Lab.

Left and previous page: Rocket Lab's Mission Control center.

Right: Launch operators on console in Mission Control.

Jane's role extends from the initial discussions with customers, through to payload design and integration, component and payload testing, and Wet Dress Rehearsals when the rocket and its payload are fully assembled together and thoroughly tested during a practice launch day. It culminates with her overseeing operations on launch day and liaising with the team at Mission Control, often while customers are present to watch their payload launch.

"It's very nerve-wracking on the day of launch, but the interesting aspect of being a Mission Manager is that you can't ever show your customer that you're nervous about something," she said.

Ten Thousand Channels of Data Coming at You

Michael Pearson, Director—Test and Launch Operations, has been on console for every Electron launch to date, since he was shoulder-tapped to play a lead role at Mission Control for the maiden flight, "It's a Test."

"My first thought was, 'Oh shit, that's the big time, that's really scary stuff. Am I comfortable doing that? Am I competent enough to do that?'" he reflected.

The Electron rocket design hasn't fundamentally changed significantly since the first launch in 2017. But the methods of interacting with the rocket when it's on the launchpad and in flight have become more streamlined and automated.

"There's less button-clicking now," said Michael. "We've added functionality to the vehicle that provides greater fidelity of telemetry coming down and real-time monitoring and data-review capabilities. You've got ten thousand channels of data coming at you. How can a human understand that? We've built tools to help us with that stuff."

His role covers preparing and running readiness reviews for rehearsals and launch activities. Flight six, "That's a Funny Looking Cactus," is one launch in particular that has stayed with him.

"I remember sitting on the floor in this factory at three in the morning with Pete, just talking through whether what we were seeing in this data was actually an issue or not," he said.

Ultimately, the team decided the rocket and payload were ready for launch. But with Michael responsible for giving the final 'Go, No-Go' approach on launch day, any doubts must be put to bed.

"That launch day was nerve-wracking, but it went off without a hitch. It shows the attitude that we have. Any little thing in the data that looks a bit odd, we think about it, and we check it out."

Despite overseeing more than sixty launches, and dozens more launch attempts that were scrubbed, he said each one elicited the same feeling. "I wouldn't say I'm any less stressed out after fifty-three launches than I was after the first. It's probably a sickness. I should take one off at least, I think."

As Team Lead, Integrated Operations Team X, Julia Wembacher leads the team that runs final pre-launch tests of Electron, and oversees the actual launch. She regularly directs a twenty-strong team at Mission Control, each person with a dedicated role, eyes glued to a screen displaying data feeds from the rocket and launch range.

For stack testing, the operation of firing up Electron's engines like it's about to launch but while keeping the rocket held down on the test mount, Julia likes to start by laying her hand on the rocket's carbon fiber shell, a ritual that represents the personal connection she feels to each launch vehicle.

On launch day itself, a further battery of tests is carried out. "Once we are happy with the dry checkouts, we commit to propellant fills," she said. "We first fill the tank with kerosene, and then we evacuate the launch site and start filling with cryogenic liquid oxygen."

The final twenty minutes before T-minus zero, the moment the rocket is expected to launch, is a strangely quiet time.

"There's no chatter anymore, you're fully focused, and your heart is racing," Julia said.

The tension is greatest during the "three, two, one" final countdown, said Jane. "There's a three-second gap between when the engines ignite and the rocket lifts off the pad. No matter how many times I watch it, that feels like the longest three seconds of my life."

You Need to Move!

As Mission Control staff oversee the launch from Auckland or Wallops Island, launch range safety experts are scanning for any hazards that might lead to a launch being scrubbed at the last minute—or second.

James Dreyer, Team Lead—Launch Safety Operations, spent fifteen years working on superyachts in the Mediterranean and Caribbean before joining Rocket Lab in 2018, just prior to Electron's third launch for the "It's Business Time" mission at LC-1.

His marine safety background was perfect preparation for Rocket Lab, given the importance of avoiding ocean-based hazards. The complex certification, compliance, and training requirements were also familiar to him from the superyacht world.

James has one of the most outwardly focused roles at Rocket Lab, such is the sheer number of stakeholders that need to be engaged to ensure the launch range is safe and secure.

Maritime New Zealand and the Civil Aviation Authority, the key regulatory bodies responsible for safety in New Zealand's sea and air realms, are key contacts. But on any given day, James will be meeting with local iwi representatives who own the land LC-1 is based on, farmers, fishermen, airline companies, and emergency services.

Rocket Lab has established infrastructure at its Mahia site, including weather balloon launching facilities and virtual beacons that alert ships to impending launches. Its notification systems for launch events have been recognized by international maritime bodies as being best practice.

But to avoid boats motoring into the range zone on launch day, sometimes a more informal approach is needed.

"Sometimes a fishing trawler has been out there for two weeks getting bashed around in pretty average weather, the crew working hard to make ends meet," James acknowledged.

An Impressive Safety Record

At a government-controlled launch site like California's US Vandenberg Space Force Base, twelve to fifteen people are likely to be assigned to launch range safety. At LC-1, it's usually a team of two, with a third occasionally added if there's a fishing competition on the calendar, or a public holiday sees amateur boaties take to the water.

Meteorologists from weather forecasting service MetService are contracted in for every launch to advise on weather conditions. Adverse weather conditions are the most common cause for a launch being scrubbed.

"It's a big deal if we scrub," said James. "Touch wood, since flight three, there's never been a scrub because of range safety, so I'd like to continue that track record."

His role has extended to overseeing safety during the recovery of first-stage Electron boosters from the ocean hundreds of miles out to sea, and the pioneering efforts to catch boosters with a helicopter.

"We are out there and hear the sonic boom of the first stage reentering. Then we see it come down under its parachute and pull up next to it in a little rigid inflatable boat in the middle of the ocean."

The spent booster is usually still hot to touch.

"On the Chatham Rise there might be sixteen-foot (5 m) swells and thirty-five knots of wind. You are dealing with something as fragile as an America's Cup yacht and thinking, 'How are we going to get this thing out of the water?'"

With Rocket Lab since flight four and on console from flight eight, Julia's first Electron anomaly was flight thirteen "Pics Or It Didn't Happen," which failed during the second-stage burn, and which was also the company's first Electron failure.

"The first one took away all the confidence as an engineer," she reflected. "I felt like, 'Yes it can happen to us as well.'"

When the going gets tough at Rocket Lab, the team comes together. "You don't do anything alone at Rocket Lab," Julia said. "They'll come in the middle of the night to help you. They'll jump on a call. Pete will jump on a call."

"We all go through the wars together," Michael agreed. "We travel together to the launch ranges. We sit on the console next to each other almost every day. We are more family than colleagues."

"And all of a sudden, some random person calls them up on the radio and says, 'Sorry, we're about to launch a rocket, you need to move!'"

— James Dreyer

You've got ten thousand channels of data coming at you. How can a human understand that? We've built tools to help us with that stuff."

— Michael Pearson

Left: The mission team watch an Electron rocket blast off.

"The tension is greatest during the 'three, two, one' final countdown . . . there's a three-second gap between when you say, 'launch' and when it ignites on the pad. No matter how many times I watch it, that feels like the longest three seconds of my life."

— Jane McNicoll

Countdown to T-Minus Zero

On launch day at Rocket Lab, a carefully choreographed series of events unfolds to ensure a successful mission. Here's an overview of the key procedures involved at Mission Control and the launch zone for an important Electron rocket launch.

→ Mission Control—Pre-launch Preparations

- → Operators are on console in Mission and Range Control centers from around T-minus ten hours initiating and monitoring fuel systems, weather forecasting, propulsion systems and more. Simultaneously, pad technicians at the launch site are on the ground preparing the rocket for lift-off.
- → The Mission Control team conducts a series of comprehensive 'Go and No-Go' polls through the countdown, with the final one typically held at T-minus twelve minutes.
- → Each station reports their status, including stage, avionics, guidance navigation and control, vehicle operations, ground control, range safety, and others.
- → The Launch Director confirms all systems are green and locks the auto sequence for launch.

-3

→ Countdown Management

- → At T-minus two minutes, the rocket's onboard flight computer takes control of the countdown.
- → Mission Control monitors critical systems and provides updates on key milestones.
- → The team oversees the final propellant loading, with liquid oxygen loading completed around T-minus one minute thirty seconds.

→ Launch and Flight Monitoring

- → The Launch Director conducts the final ten-second countdown.
- → The engines ignite at T-minus two seconds. Hold-down clamps affix the rocket to the pad for those final two seconds before lift-off, allowing onboard flight computers to conduct a final check that all systems are healthy, before releasing the rocket for flight. It's possible for the computer to diagnose a potential issue at this point and shut down the engines before lift-off.
- → In flight, the rocket is entirely autonomous. Mission Control operators simply track the rocket's performance by monitoring trajectory, speed, fuel consumption, and altitude throughout the flight.
- → The journey from lift-off to orbit is typically around nine minutes, with payloads usually deployed within an hour of lift-off once Electron's Kick Stage has positioned the payloads precisely where they need to be released.

-2 -1

ELECTRON

A New Twist on Electron

HYPERSONIC
ACCELERATOR
SUBORBITAL
TEST
ELECTRON
↓

HASTE

By 2022, Electron was regularly delivering satellite payloads into Low Earth Orbit (LEO). But Rocket Lab's military customers proposed a slightly different use of the tried-and-tested launch vehicle—suborbital technology test missions at hypersonic speeds.

Geopolitical tensions have spurred development of hypersonic capabilities. China and Russia have already demonstrated operational hypersonic weapons capabilities, which pose a significant threat to US national security.

To be able to compete, the US Department of Defense needs to urgently prototype and test their own hypersonic technology. There are heat problems, structural integrity problems, and radar tracking problems that occur when you're moving faster than the speed of sound, and all of those issues need to be ironed out in the concept and development phase. But for a long time, the only way to test was by using wind tunnels or sounding rockets that generated only fractions of a second of relevant hypersonic data.

In the minds of the US Pentagon's top officials, that pace of technology development needs to move faster. They wanted a higher cadence of tests that can move at hypersonic speeds, and with Rocket Lab's Electron rocket—highly configurable and maneuverable during launch, already flying regularly, and operating at an affordable price—they saw an opportunity.

HASTE—Rocket Lab's Hypersonic Accelerator Suborbital Test Electron—was born. Derived from Electron, HASTE uses Electron's proven carbon fiber composite structure and 3D-printed Rutherford engines, but has a larger payload capacity of up to 1,540 pounds (700 kg), since its mostly flies at suborbital altitudes. With its highly maneuverable engines, HASTE can fly many of the trajectories that the US is wanting to test—from direct inject trajectory, to boost-glide, parabolic, and reentry trajectories—and meet the payload release conditions to satisfy mission-specific objectives.

HASTE missions operate from Rocket Lab's LC-2 at NASA's Wallops Flight Facility in Virginia, ensuring rapid deployment capabilities and the ability to undertake sensitive missions.

1,540 POUNDS (700 KG) PAYLOAD MASS
HASTE employs the same innovative carbon composite structure and 3D-printed Rutherford engines as Electron but has a modified third stage for suborbital payload deployment, a larger payload capacity of up to 1,540 pounds (700 kg), and options for tailored fairings to accommodate larger payloads.

MISSIONS TO DATE	**"Scout's Arrow" (June 17, 2023)** The inaugural HASTE mission launched under the Multi-Service Advanced Capability Hypersonics Test Bed (MACH-TB) program for the US Department of Defense. The objective was to validate hypersonic-glide body components and gather critical test data.	**"HASTE A La Vista" (November 24, 2024)** Conducted under the MACH-TB program to advance hypersonic technology research for the US Department of Defense. Featured experimental instrumentation and demonstrated high-speed testing capabilities on an accelerated timeline.
UNDISCLOSED	A third hypersonic test mission on HASTE under the MACH-TB program in support of technology maturation.	
UPCOMING MISSIONS	**"HyCat" (2025)** A collaboration with the Defense Innovation Unit (DIU) to deploy Hypersonix's scramjet-powered DART AE vehicle during ascent. The aim is to showcase HASTE's "direct inject" capability within Earth's atmosphere at speeds of up to Mach 7.	**"MACH-TB 2.0" (2026–)** HASTE has been selected to launch for the next round of the MACH-TB program, with the first mission awarded under MACH-TB 2.0 scheduled to launch from Rocket Lab Launch Complex 2 in Virginia in 2026. There is the possibility of multiple HASTE launches being awarded under the five-year contract, which has a total value of $1.45 billion.

ROCKET
GRINDING
SOUND PROOF 02
RESIN MIXING BAY

ROCKETLAB

The Big Time —Going Public on the Nasdaq

"We looked around to the rest of the space industry that had also pursued SPACs, and there was just destruction, carnage, and carcasses everywhere."

— Sir Peter Beck

By the end of 2020, Rocket Lab had seventeen successful Electron missions behind it, a fast-growing Space Systems business and a healthy collection of deals in the pipeline. But Peter Beck was looking to take the company to the next level, and for that, he'd need cash, more than could be easily raised in another venture funding round.

In March 2021, Rocket Lab announced the Neutron program. The company would develop a partially reusable rocket to deliver payloads to Low Earth Orbit and beyond, opening up competition in the mid-lift market dominated by SpaceX.

"We always wanted to do this medium-scale launch vehicle, and we wanted to scale the Space Systems business and ultimately get into applications," said Peter. "That takes a lot of capital, a tremendous amount of capital."

Peter and the Rocket Lab board saw an opportunity to raise money relatively quickly by taking Rocket Lab public by way of a special purpose acquisition company (SPAC), rather than a traditional initial public offering (IPO).

SPACs had become popular in the US in 2019. In 2020, 247 SPACs raised over $80 billion in investment. Often referred to as a "blank check company," SPACs are listed companies that raise money and then go on the hunt for a private company to merge with, taking it public in the process. In the low-interest rate environment following the pandemic, investors were hungry for deals, fueling a flurry of interest in SPACs, which were seen as a faster and more efficient way to go public.

In 2021, during the peak of the SPAC boom, 613 companies raised $162 billion. Rocket Lab was among them when it went public on the Nasdaq in August 2021, with the listed company Vector Acquisition.

Rocket Lab's Coffers Grow

The listing valued Rocket Lab at $4.8 billion, and raised $777 million in gross cash for the company, filling its coffers and giving it the capital to forge ahead with Neutron, which was expected to cost up to $300 million to develop. At the time, Rocket Lab had 500 staff. The listing would fund a rapid period of headcount growth at Rocket Lab, the most intense period of hiring in the company's history.

A Nasdaq listing is seen as the pinnacle of investor endorsement for technology companies, often the ultimate exit for founders eager to reap the rewards of building a fast-growing company.

But Peter saw it as something different. Sure, it generated the cash needed to realize Rocket Lab's ambitions, Neutron included. But it gave Rocket Lab better prospects of enduring for decades to come.

"I don't want this to be the Peter Beck show," he said. "If I die and this company dies, then that is absolute failure, failure at the highest level."

Ringing the Nasdaq Bell—Remotely!

Most chief executives of newly listed Nasdaq companies get to celebrate the event by ringing the opening bell at Nasdaq's headquarters in New York, followed by the obligatory round of interviews on CNBC and Bloomberg TV.

But Peter wasn't in New York that day. With a new wave of Covid cases spreading in August 2021, he didn't want to run the risk of being stranded in the US as New Zealand was plunged back into lockdown.

It was a wise move. At midnight on August 17, 2021, New Zealand entered Alert Level 4, requiring people to once again stay at home. Auckland would remain in lockdown for the next seven weeks.

Peter commissioned a local signwriter to create a mock-up of the Nasdaq plinth, and gathered staff on the factory floor in Auckland to record a video to celebrate going public. Then the team went home to lockdown and the video went to Nasdaq, where it aired on August 25 when Rocket Lab (RKLB) shares started trading.

It wasn't long before the SPAC bubble burst, with companies failing to live up to investors' expectations quickly punished with depressed share prices. Rocket Lab was caught up in the fallout.

"We put all of our numbers out, and said what we were going to do. To this day, we've largely met those targets," said Peter. "We looked around to the rest of the space industry that had also pursued SPACs, and there was just destruction, carnage, and carcasses everywhere."

Rocket Lab, in contrast, was fundamentally a solid company with growing revenue and credible plans for expansion. But it wasn't immune to SPAC backlash. Rocket Lab stock opened at US$11.58 on its first day of trading. By early September, the stock was down 15 percent, before jumping 20 percent soon after.

The stock price remained volatile before the toxic sentiment around SPACs and the tougher economic climate driven by inflation and high interest rates saw RKLB stock languish around the US$5 mark for much of the period between 2022 and 2024.

"It's always depressing when your stock price is going down," Peter admitted. "It was a few years of walking through rocks. But that's just fine, no big deal at all. You've got to keep your eye on the longer-term prize."

In late 2024, Rocket Lab shares enjoyed a spectacular rally as the market absorbed news of its progress on developing Neutron, and the launch and Space Systems divisions of the business enjoyed strong growth. Rocket Lab shares have since soared more than 500 percent over 2024 to 2025.

Running a listed company introduced a new dynamic—intensive scrutiny of Rocket Lab's operations, and the financial reporting requirements that go with being a publicly listed company.

"I don't sleep very well ever," said Peter. "It's just worrying about different things, and now that Rocket Lab is a publicly traded company it's a whole different set of worries."

While some space tech companies that listed due to a SPAC merger failed, Peter said the flurry of investment and attention did boost the commercial space market. "I generally believe that, even into silly ideas, great people were developed, some new technologies were developed. It wasn't a total loss in most cases," he told the SmallSat Symposium via a video keynote in February 2025.

He valued the financial discipline quarterly earnings announcements demanded. "Any entrepreneurial CEO will tell you that running a public company is hard and not that much fun. But it forces me to focus on profitability, which ultimately drives to my longer-term goal of endurance and having a longer-lasting impact," he said.

This view contrasts with that of Elon Musk, who has stated a desire to keep SpaceX as a private company as the short-term demands of the market conflict with his long-term goals for the company. Instead, SpaceX has raised money in private fund-raising rounds, including most recently in December 2024, in a round that reportedly valued the company at $350 billion.

SpaceX staying private served to increase investor interest in Rocket Lab, the only listed space industry company regularly launching payloads to orbit. For years, Rocket Lab has livestreamed its launches, been active on social media, and active in industry events and conferences to explain its approach to launching rockets and building satellites.

But with the heightened financial scrutiny that came with a Nasdaq listing, the company had to change how it communicated its objectives, and issued updates on progress towards them.

"It's not an industry for the fainthearted," said Morgan Connaughton, Vice President—Marketing and Communications. "All of a sudden you're talking to people who are excited by rockets and excited by satellites, but probably don't understand them in any great detail, and certainly don't understand necessarily the risk involved in this industry."

Owning launchpads in New Zealand gave Rocket Lab the freedom to set and adjust its own launch schedule in an industry where everything from weather to last-minute glitches with payloads can push out a launch at the last moment.

"Our whole business model is predicated on giving customers flexibility such that, if they need an extra week for spacecraft readiness, they haven't missed their rideshare," Morgan explained. "There were times we'd push a launch into the next quarter and the market would react negatively, but the ability to hold a launch by a few days is actually a luxury and a core part of what makes Electron attractive to customers. So we had to do a whole education piece around the fact that by the time we get to launch day, we've collected 90 percent of the revenue from that launch."

SpaceX remaining a private company has seen investors keen to back space industry ventures faced with slim pickings among listed companies. Rocket Lab, with its impressive track record of launches and vertically integrated business model, has therefore attracted growing interest from analysts and the investment community. By the end of June 2025, Rocket Lab had a $16 billion market capitalization making it one of the most valuable space companies in the world.

Previous page: Electron HASTE rockets at Rocket Lab's Auckland production complex, New Zealand.

Left and overleaf: Peter Beck speaks with the Rocket Lab team in celebration of the company's then upcoming debut on the Nasdaq stock exchange in August 2021.

ROCKET LAB
SOUND PROOF 02
RESIN MIXING BAY
CNC PLOTTER

Nasdaq

PART III

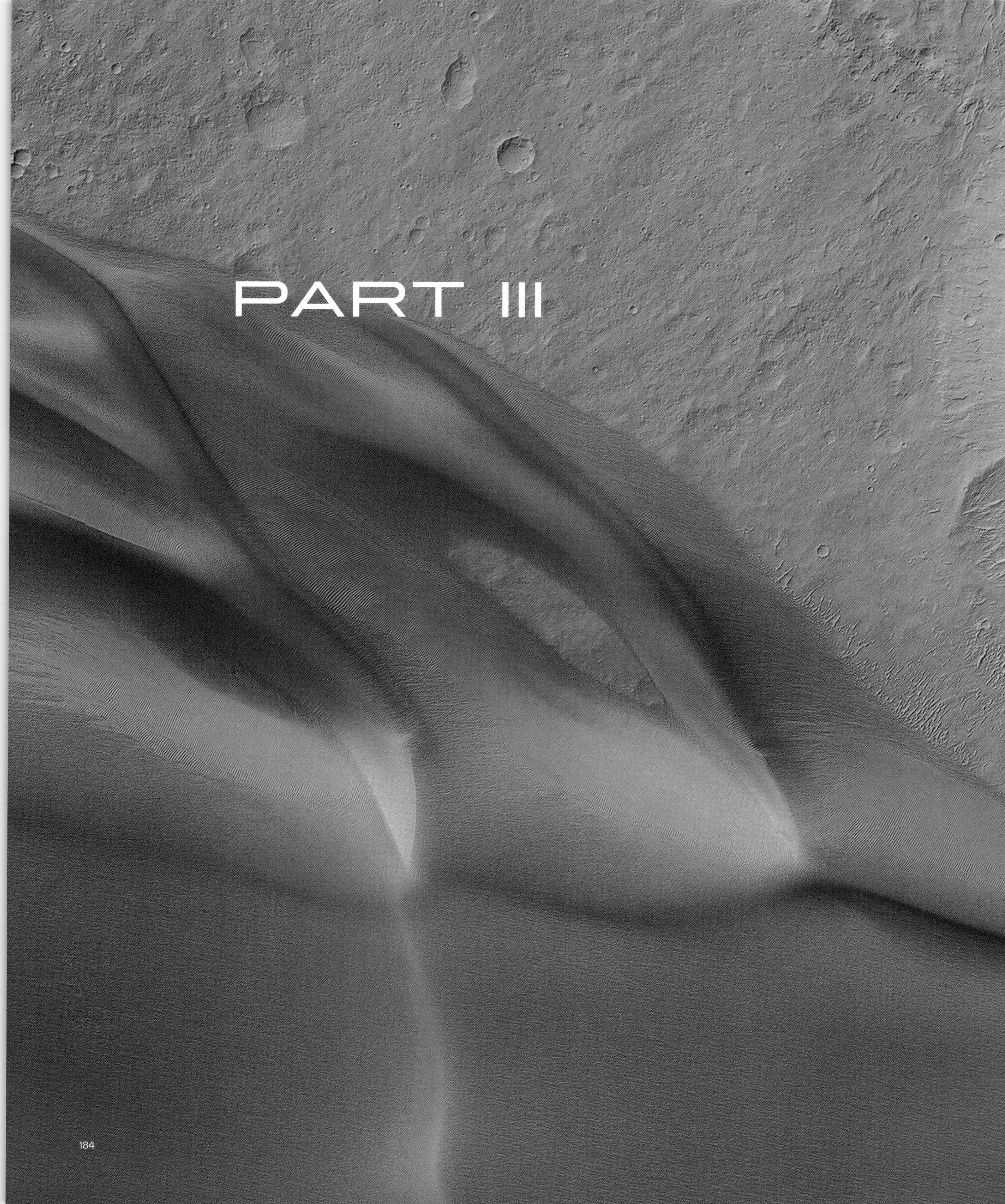

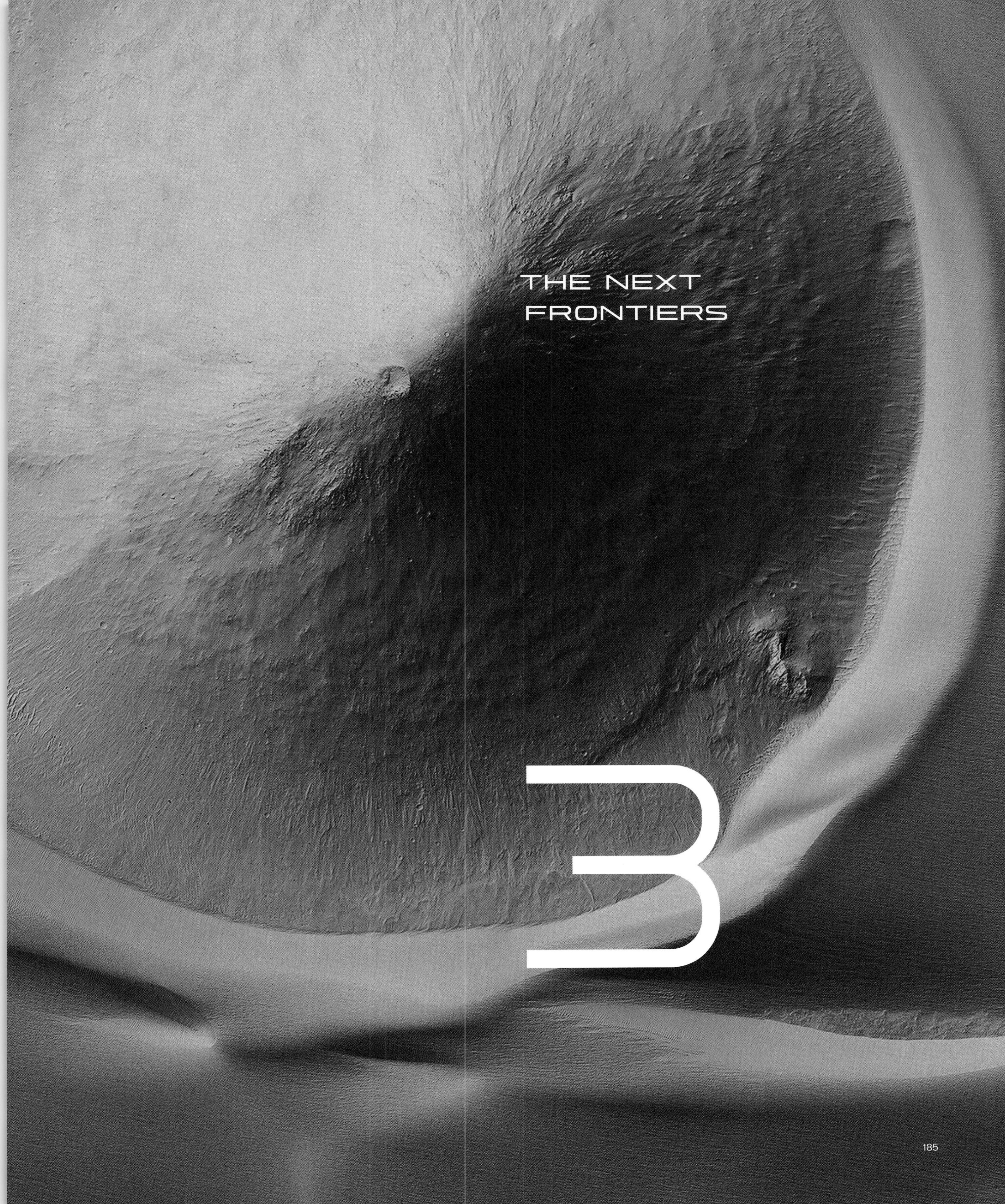

3

THE NEXT FRONTIERS

Infrastructure in Space

With Electron and Neutron capturing most of the headlines written about Rocket Lab, it's easy to overlook the fact that the company is also in the business of building satellites and their components that are propelled into orbit by those rockets and other companies' launch vehicles.

Richard French, Vice President—Business Development and Strategy, Space Systems, Rocket Lab.

In fact, Rocket Lab's Space Systems division currently contributes around 70 percent of the company's revenue. Peter Beck's initial focus was to build a reliable, low-cost launch vehicle, which became Electron.

But he recognized early on the need to expand into spacecraft and satellite components to provide end-to-end space solutions.

The Electron Kick Stage that carried the payload into orbit during flight two, "Still Testing," in January 2018, had recessed fittings for solar panels. It had been designed to be turned into a functioning satellite.

"That was always part of the master plan," said Peter. "The moment we got Electron humming, around 2019, we moved into Space Systems and satellites."

The Kick Stage, designed to precisely position payloads in orbit, became the basis of the Photon spacecraft, which was officially announced in April 2019. But making Space Systems an integral part of Rocket Lab's offering required a two-pronged approach—developing technology in-house and acquiring companies already building sought-after space tech.

Peter's first foray into acquisitions led him to Doug Sinclair, the founder of Canadian satellite hardware maker Sinclair Interplanetary, initially just to procure key components for Photon—namely star trackers and reaction wheels.

Star trackers are optical sensors that help satellites maintain their orientation and location in space. Reaction wheels are a type of electric motor attached to a flywheel that allow satellites to be repositioned while in orbit.

Sinclair Interplanetary was considered one of the world's best producers of both. But when Peter put in a large order with Sinclair in mid-2019, he received an unexpected response. Sinclair explained he had just closed his order book, and the parts would take nine to twelve months to deliver to Rocket Lab.

"That wasn't going to work," said Peter. "We only had twelve months to build a satellite."

Rocket Lab's Space Systems Business Takes Shape

Until recently, the space industry relied on a cottage industry of small, specialist producers building quality hardware, but in small volumes.

"Whenever we'd turn up to those guys and ask for a thousand units, they'd have a mental explosion. It just didn't work," Peter said.

So he offered to buy out Sinclair Interplanetary, merge it into Rocket Lab, and ramp up production. He said he was "super lucky" that Sinclair agreed to sell his business.

"Doug and his team used to do a hundred and fifty reaction wheels a year. This year alone we'll do over 2,000 reaction wheels," he said.

Sinclair Interplanetary would be the first of four major acquisitions made between 2019 and 2022 that gave critical mass to the Space Systems division, bringing in top talent and specialist lines of hardware, from satellite components to solar cells, avionics, and all-important flight and ground software.

Richard French, who joined Rocket Lab in 2019 as the company's most senior Space Systems expert, recalls the early days with a mix of humility and excitement.

Coming from a distinguished NASA and JPL background, including Mars rover missions and synthetic aperture radar satellites, Richard was drawn to Rocket Lab's shared passion for pushing boundaries. He recounts his first conversations with Peter Beck:

"I asked Pete, 'So you're gonna let me build satellites?' He said, 'Yeah.' I asked, 'What about NASA science missions?' He said, 'Absolutely.' And then I asked, 'How about planetary missions?' And he said, 'Are you kidding? That's one of my passions.' And I said, 'Alright, sign me up.'"

At that time, Rocket Lab was still early in converting the Electron rocket's Kick Stage into what would become the Photon satellite platform. Richard, who now serves as Vice President—Business Development & Strategy for Rocket Lab's Space Systems division, describes the team back then as "six guys soldering printed circuit boards in the back room," underscoring how nascent the Space Systems effort was even as the company was gaining launch experience.

CAPSTONE as a Catalyst

The launch of the Photon platform and the landmark CAPSTONE mission to the Moon in 2022 marked a turning point.

"We asked ourselves, what kind of company do we want to be? One that solves hard problems was the answer. All the easy stuff has been done. There are only hard problems remaining," said Richard.

"CAPSTONE was a great example of a hard problem that rallied the company around a shared vision."

The CAPSTONE mission required improvements to rocket engines, stretching Electron's second stage and lightening the launch vehicle to enable the team to deliver NASA's precious payload into lunar orbit. All of this innovation came before Rocket Lab was able to benefit from expertise brought into the fold by way of acquisitions. But it demonstrated the desire to push the boundaries of engineering in order to meet its customers' needs.

Richard was integral to helping Peter and Rocket Lab's executive team identify and pursue those acquisitions, which Richard notes were not just about growing the business but about gaining discriminating technologies critical for demanding missions—precision attitude control, radiation tolerance, propulsion, and more.

Previous page: An interior view of Rocket Lab's Lunar Photon spacecraft for the NASA CAPSTONE mission to the Moon.

This page: Rocket Lab's 12Nms Reaction Wheel.

"That was always part of the master plan. The moment we got Electron humming, around 2019, we moved into Space Systems and satellites."

— Sir Peter Beck

P/N: ST-16RT
S/N: 002299

349

Reaction Wheels and Star Trackers

Rocket Lab's reaction wheels and star trackers are at the heart of modern satellite attitude control and navigation, trusted across the global space industry for their precision and reliability.

With more than 240 reaction wheels currently operating in orbit, these high-reliability, vacuum-lubricated devices are essential for picosatellites, nanosatellites, and microsats, offering precise three-axis control without the need for propellant. Rocket Lab's star trackers, including the new mass-producible ST-16HV model, provide unmatched accuracy in determining a spacecraft's orientation by referencing star fields, making them indispensable for both commercial and government satellite constellations.

Manufactured for rapid deployment and high-volume constellations, these components are integrated into a wide variety of missions—from one-off technology demonstrators to large-scale Earth observation and communications fleets.

Their proven heritage and ongoing innovation have cemented Rocket Lab as a leading supplier of flight-proven, affordable, and customizable satellite components for the expanding space industry.

Previous pages: Under production: Rocket Lab's high-performance star tracker for small spacecraft.

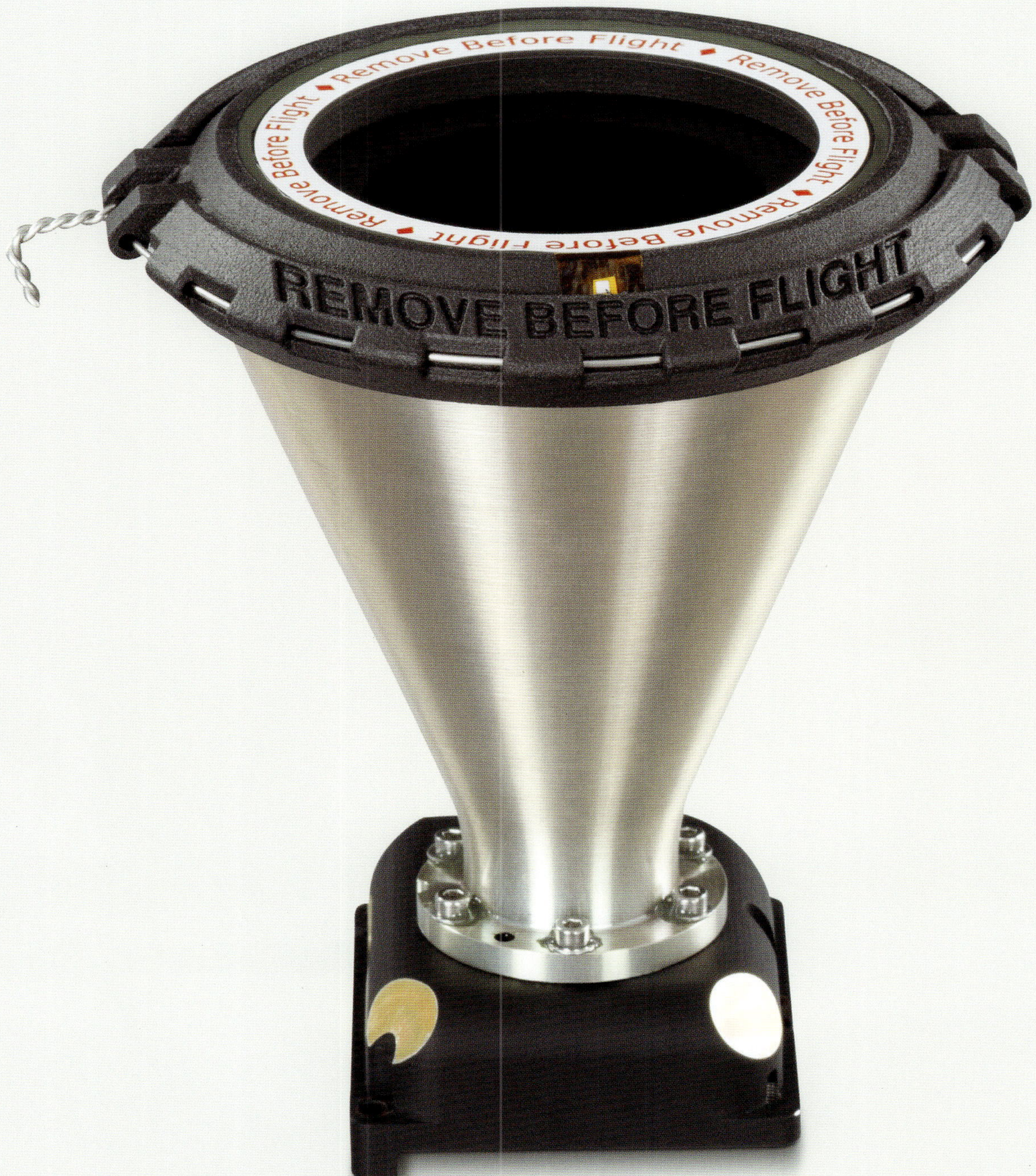

Left: Rocket Lab's 30 mNms reaction wheel, optimized for use on a 6U to 12U CubeSat.

This page: Rocket Lab's ST-16RT2 star tracker.

Left and right: Rocket Lab's solar cells in production in Albuquerque, New Mexico, US.

"Space cells use compound semiconductors like gallium arsenide, which are up to 100 times more expensive but 30 percent more efficient and radiation hardened. This efficiency enables satellites to be lighter and more powerful, supporting more channels and connectivity."

— Dr. Brad Clevenger, Vice President—Space Systems, Rocket Lab

Dr. Brad Clevenger, Rocket Lab's Vice President—Space Systems, arrived at Rocket Lab in 2022 when the company he ran, Albuquerque, New Mexico-based SolAero Holdings, was brought into the fold. An analytical chemist by training, Brad started his career at the Naval Nuclear Laboratory, researching the most efficient ways to convert nuclear power into electricity to power warships, aircraft carriers, and submarines.

Brad ended up jumping ship to the private sector, to a company called EMCORE, where he was put to work on compound semiconductor materials.

"All that stuff we were developing at the time was kind of esoteric, but within five years, compound semiconductor materials that were being used to manufacture cell phones and other miniaturized electronics. The materials, gallium arsenide and indium phosphide among them, proved particularly well suited for use in solar panels to power spacecraft," he said.

"Space solar power is fundamentally different from terrestrial solar, typically made of silicon. Space cells use compound semiconductors like gallium arsenide, which are up to 100 times more expensive but 30 percent more efficient and radiation hardened. This efficiency enables satellites to be lighter and more powerful, supporting more channels and connectivity.

"Within three years of building a manufacturing plant in Albuquerque, more than 90 percent of the solar power going into space was based on compound semiconductors," he said.

Joining Rocket Lab as part of the SolAero acquisition, Brad was made an offer he couldn't refuse—heading up the burgeoning Space Systems division and integrating established space companies into Rocket Lab's fast-paced culture.

"Each of those companies was older and more established than Rocket Lab, with baggage—good products, great heritage, but culturally not always the best fit," he said.

"Pete has a very clear vision for how the company operates, and not everyone made it through the transition. But the products and smart people were worth adapting for."

The solar power capability that came to Rocket Lab in the form of SolAero technology and expertise is today perhaps the most striking example of the reach Rocket Lab now has across the space industry.

"Rocket Lab powers virtually every national security and civil space mission that Boeing doesn't manufacture. We supplied 80 percent of the solar power for the OneWeb generation one constellation," Richard pointed out.

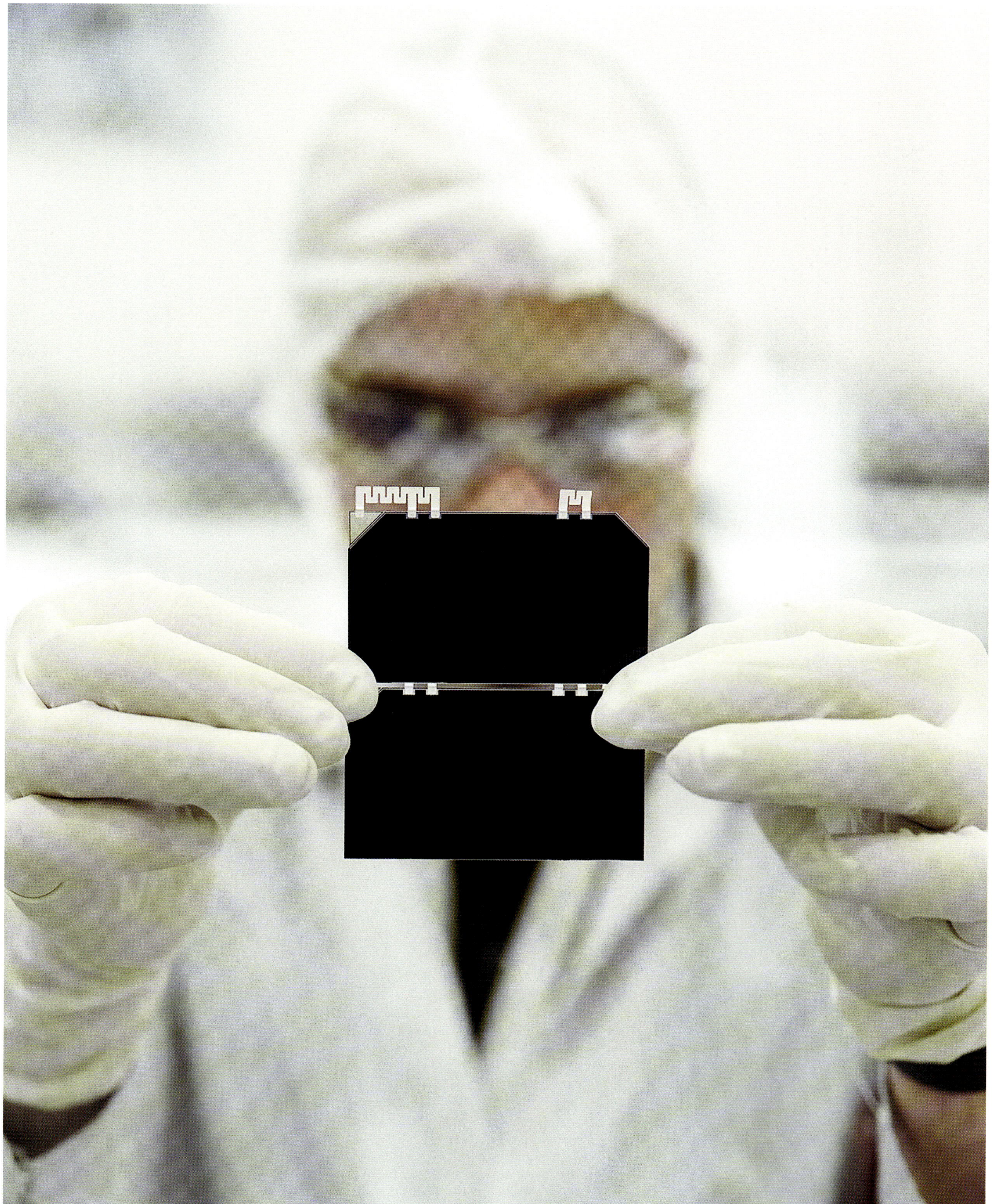

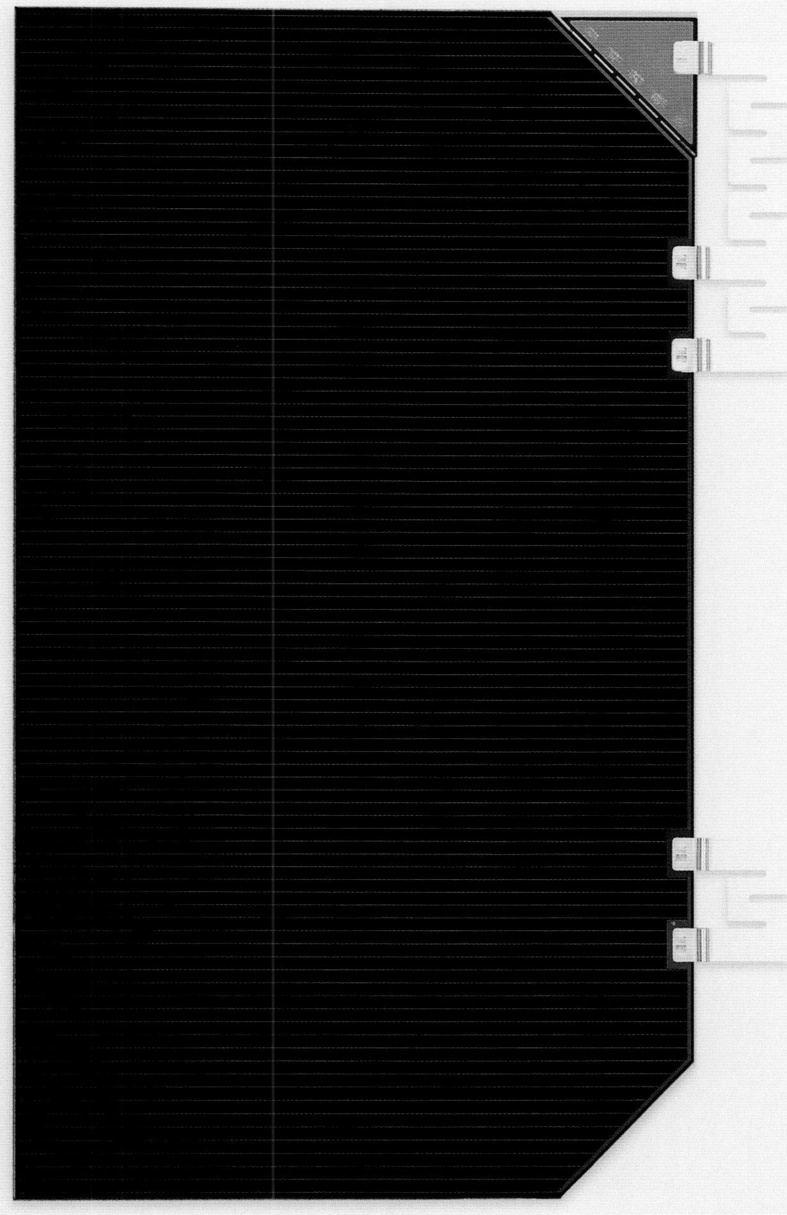

These pages: Rocket Lab's high-efficiency solar cells.

Solar Power, Rocket Speed

Rocket Lab's Space Systems division is redefining satellite power with its advanced, customizable solar arrays—delivering reliable, high-efficiency solutions that power everything from commercial constellations to deep space missions.

Powering satellites and spacecraft is most commonly achieved by harnessing the ever-present power of the Sun. Solar arrays have been launched into space for decades and Rocket Lab has rapidly established itself as a global leader in space-based solar power, offering a vertically integrated manufacturing capability that spans from high-efficiency solar cells to fully integrated solar arrays.

A pivotal moment in Rocket Lab's solar journey was the 2022 acquisition of SolAero, a renowned specialist in space solar technology. This strategic move enabled Rocket Lab to internalize the production of radiation-hardened, flight-proven solar cells, cementing its position as the only provider with a fully vertically integrated supply chain for space solar arrays. As a result, Rocket Lab now powers over 1,100 satellites in orbit, including flagship missions like the James Webb Space Telescope, NASA's Artemis lunar program, and commercial networks like OneWeb.

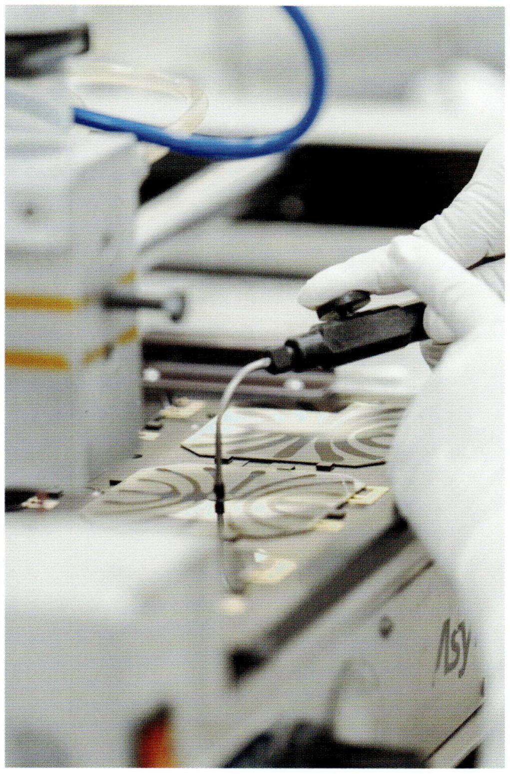

Rocket Lab's new STARRAY product line delivers customizable solar solutions, supporting satellite power needs from 100 to over 2,000 watts, with options for both rigid and flexible panels. This flexibility allows Rocket Lab to serve a diverse array of missions—from CubeSats and large constellations in Low Earth Orbit to lunar and interplanetary spacecraft—while maintaining short lead times and cost efficiency.

These pages: Rocket Lab's solar cells in production in Albuquerque, New Mexico, US.

Overleaf: Rocket Lab solar panels about to undergo environmental testing within a thermal vacuum chamber.

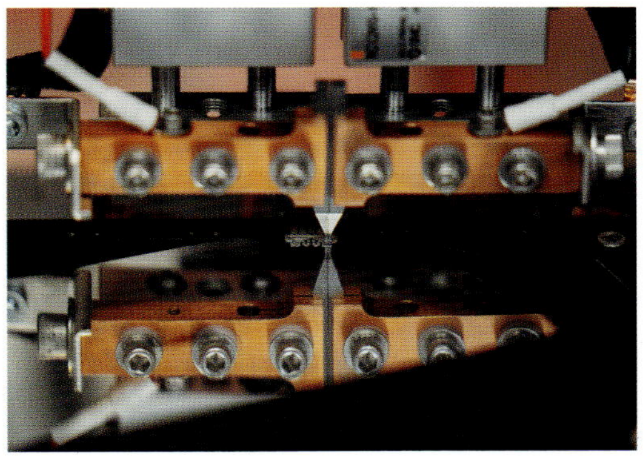

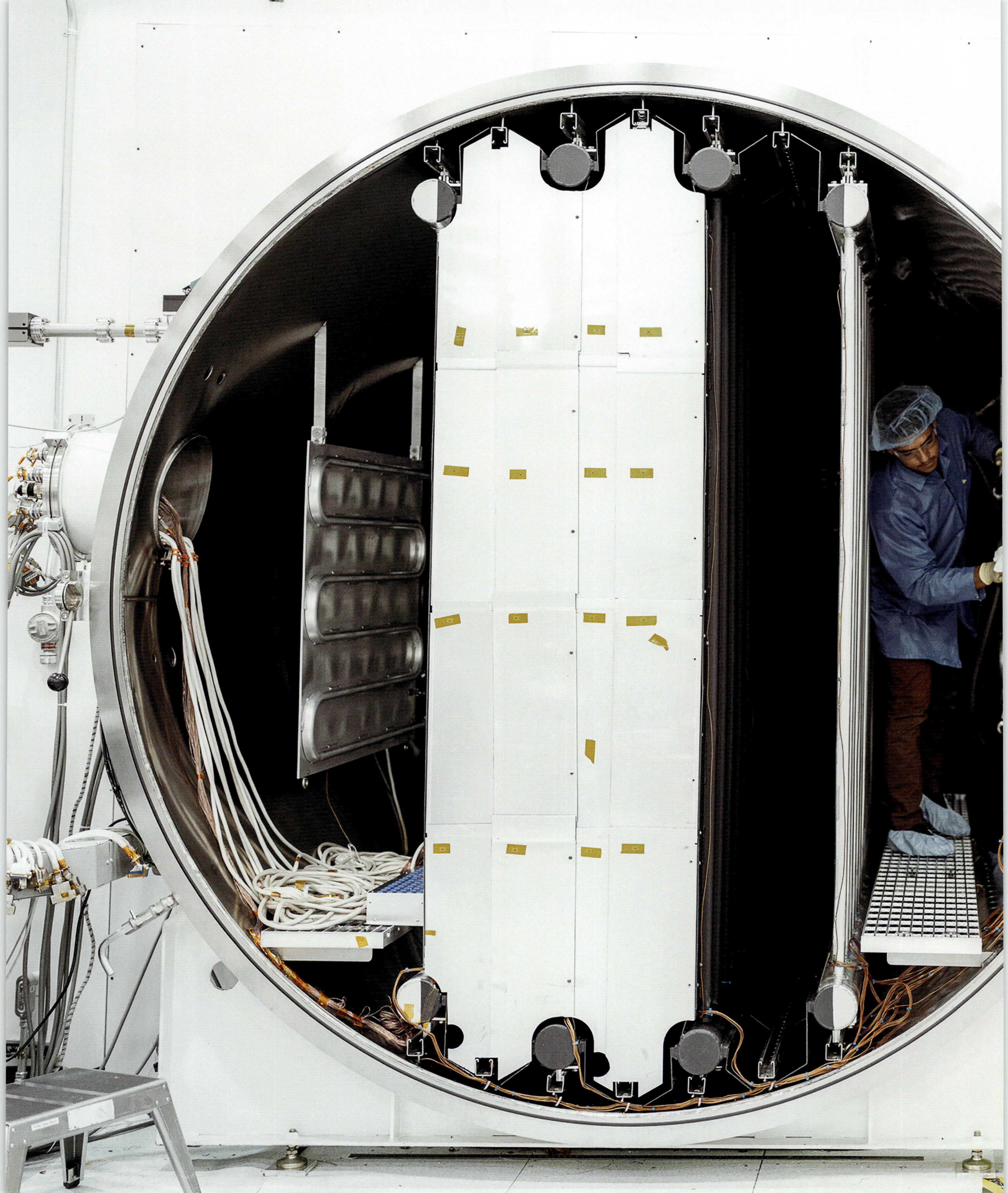

When Competitors are Customers

As Rocket Lab's Space Systems business grew, Peter's philosophy was to be as collaborative with other space industry players as possible. That extends to working with competitors such as Firefly Aerospace, a Texas-based developer of small- and medium-lift launch vehicles.

"Most Americans wouldn't change their own oil or fix their own appliances, but Kiwis do. That resourcefulness is ingrained in Rocket Lab's DNA and drives us to solve problems ourselves rather than pay more or buy from struggling suppliers."

— Dr. Brad Clevenger

Rocket Lab played a pivotal role in the success of Firefly Aerospace's Blue Ghost Mission 1 for NASA, the first fully successful commercial soft landing on the Moon. Rocket Lab's MAX Flight and MAX Ground Data Software suites provided the backbone for mission operations, delivering essential functions such as commanding, telemetry, autonomous sequencing, guidance, navigation, control, and fault protection throughout every phase—from launch to lunar touchdown.

Rocket Lab's Mission Operations Center in Littleton, Colorado worked in tandem with Firefly's Texas base, handling orbit determination, maneuver planning, command generation, and continuous health monitoring of the lander's systems.

"Firefly is one of our competitors on the launch side, and yet, they're a great customer of ours on the space system side," said Richard French, Vice President—Business Development and Strategy, Space Systems. "We were handling one of the hard problems for them, coming up with the software to help them land on the Moon."

Rocket Lab also supplied three high-efficiency photovoltaic assemblies, generating 400 W of power over 1,470 operational hours to sustain the lander's journey and surface operations. Blue Ghost touched down on the Moon on March 2, 2025 and for the best part of the next two weeks undertook numerous experiments as part of NASA's Artemis project.

In the case of BlackSky, which operates a satellite constellation offering high-resolution satellite imagery, analytics, and high-frequency monitoring, Rocket Lab has contributed to most aspects of the satellites, from solar arrays to reaction wheels, star trackers, and separation systems. BlackSky, which has built out its constellations with launches between 2019 and 2025, also uses Rocket Lab's software and the majority of its satellites have been delivered into orbit on Electron rockets.

The list of missions continues to grow, with each one demonstrating a different facet of Rocket Lab's capabilities in Space Systems.

Dr. Brad Clevenger, Vice President—Space Systems, is particularly proud of the work Rocket Lab has done with Varda Space Industries to support its on-orbit manufacturing satellites.

Rocket Lab built and operated the host satellite that delivered manufacturing capsules to orbit and successfully returned them to Earth, becoming the first commercial company to land and recover a capsule intact.

"Landing a capsule moving at 17,000 miles per hour (27,358 km/h) requires calculations to five decimal places—differences of tens or hundreds of kilometers mean success or disaster. We did it three times in a row, while others have failed," Brad said.

Then there's NASA's "ESCAPADE" Mars mission to measure plasma and magnetic fields around the Red Planet with the aim of better understanding what's driving Martian climate change.

Rocket Lab designed and built twin spacecraft called Blue and Gold to enable the mission. But it ran into a problem in the development phase. The company that was tasked with building the thin-wall titanium fuel tanks for Blue and Gold hadn't produced a tank halfway through the eighteen-month production cycle.

"So we invented a tank from scratch," said Richard.

"We 3D-printed it on a printer that we normally use for rocket engines and machined it into a tank. It was the first of its kind approved for use on the spacecraft by NASA."

Brad puts that culture of resourcefulness down to Rocket Lab's New Zealand heritage.

"Kiwis are incredibly resourceful by necessity," he said. "Most Americans wouldn't change their own oil or fix their own appliances, but Kiwis do. That resourcefulness is ingrained in Rocket Lab's DNA and drives us to solve problems ourselves rather than pay more or buy from struggling suppliers."

Engineering Excellence Plus Financial Discipline

After more than five years at Rocket Lab, Richard said the company's unique culture gives it an edge in developing Space Systems.

"Rocket Lab works its people really hard," he pointed out. "Those of us who last have gas tanks that never run empty. The company has a survival mentality—if we don't finish on time, we die. Pete's leadership is engineering-driven, our Chief Financial Officer Adam Spice's is financial discipline. They trust people and don't micromanage. That combination keeps us lean and focused."

Looking ahead, the availability of Rocket Lab's Neutron rocket will only increase the opportunities for the Space Systems division.

"Neutron lets us optimize satellites and rockets together, rather than bending to SpaceX's dominant Falcon 9 design," said Brad.

"It opens opportunities to launch larger payloads and constellations with fewer trade-offs. It also provides an alternative launch option for customers seeking competition and flexibility."

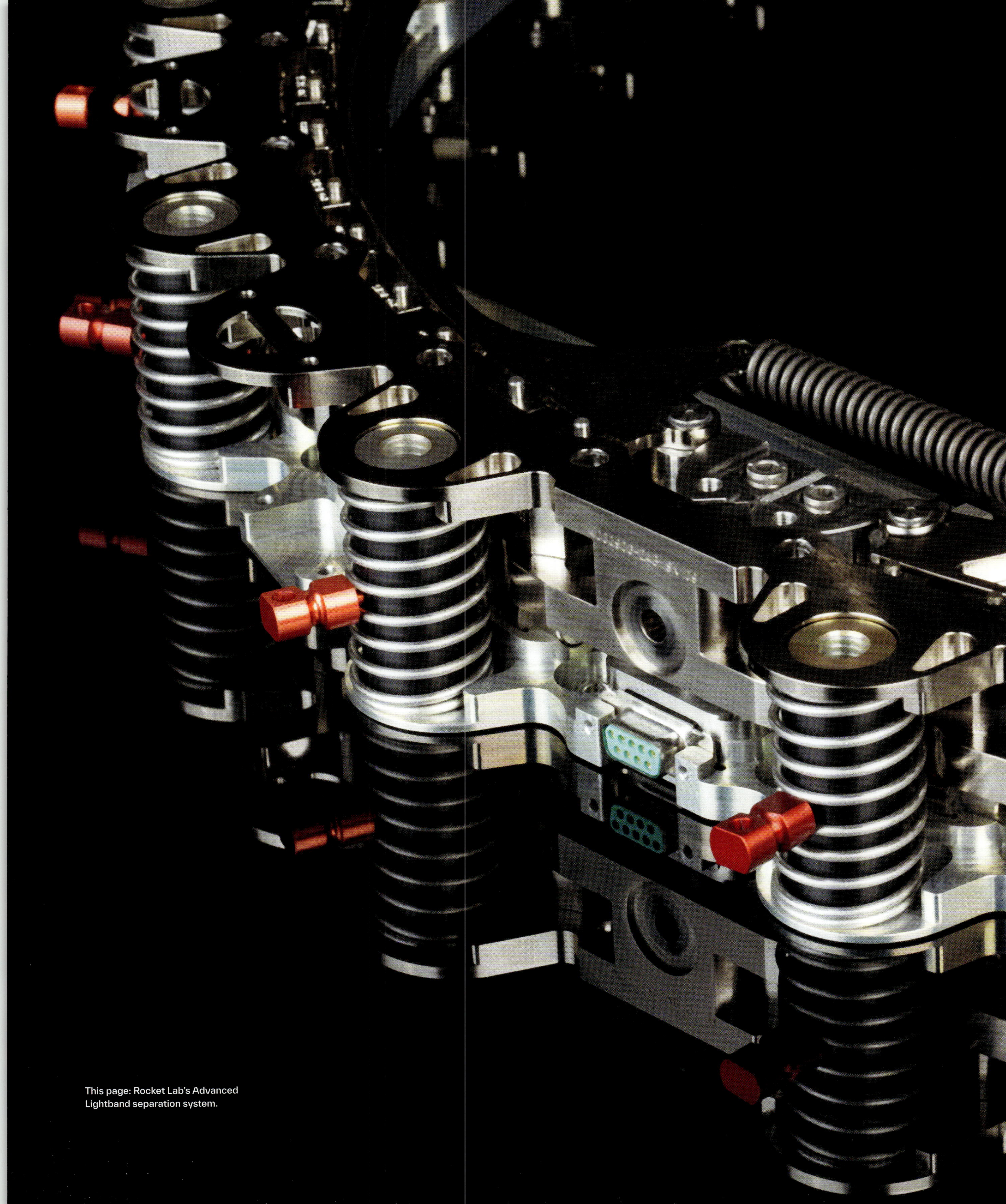

This page: Rocket Lab's Advanced Lightband separation system.

Separation Systems

Rocket Lab's separation systems are the unSung heroes behind successful satellite deployments, enabling spacecraft to safely and precisely detach from launch vehicles in orbit.

The result of more than twenty-five years of research and development, these advanced, lightweight mechanisms—including the renowned Lightband and Canisterized Satellite Dispenser (CSD)—are engineered for reliability, rapid integration, and cost-effectiveness.

Used across most active US launch vehicles and trusted by NASA and global satellite operators, Rocket Lab's separation systems support a range of missions, from small CubeSats to major defense constellations.

Their innovative, non-pyrotechnic designs minimize risk and simplify payload integration, reducing preparation time from days to minutes. Rocket Lab's separation systems have achieved a flawless 100 percent mission success rate across hundreds of deployments, making them the industry standard for dependable satellite delivery and mission assurance.

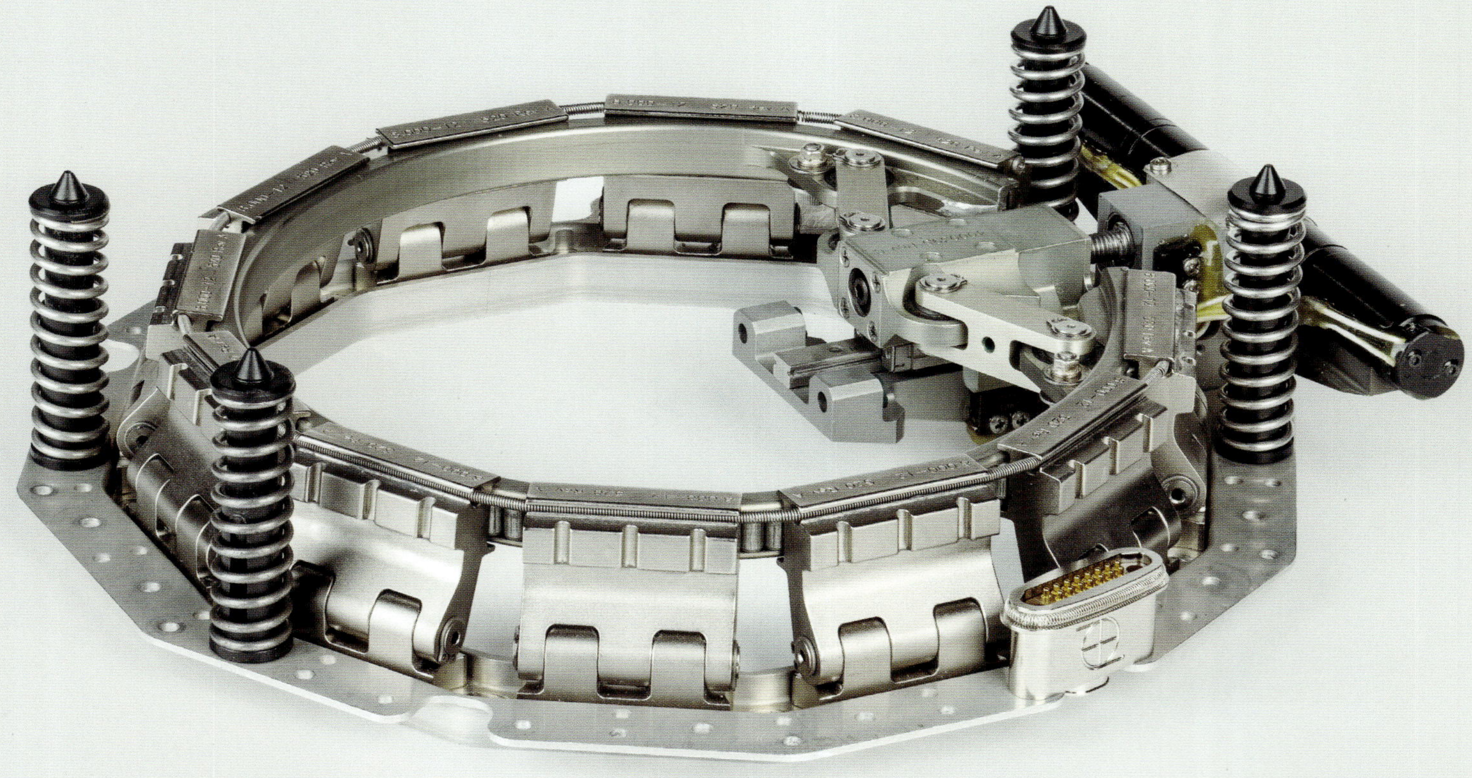

These pages: Rocket Lab's Motorized Lightband separation system.

Coding for Space

Rocket Lab's software offerings have evolved into some of the most advanced and widely adopted solutions in the space industry, powering everything from launch to deep space operations.

Its flagship MAX Flight Software, developed by Advanced Solutions Inc. and now operating on over fifty spacecraft, supports a full spectrum of mission needs—command, telemetry, sequencing, guidance, navigation, control, and fault protection—across commercial, government, and defense programs, including NASA's CAPSTONE and Firefly's Blue Ghost lunar lander.

Building on this legacy, Rocket Lab recently introduced InterMission and MAX Constellation, next-generation platforms designed for autonomous, secure, and scalable constellation management and real-time spacecraft operations. These tools enable users to monitor, command, and analyze fleets of satellites with robust encryption and customizable data workflows.

With over 250 cumulative years of on-orbit heritage, Rocket Lab's software is now a backbone for some of the world's most ambitious space missions, trusted by leading agencies and commercial operators alike.

These pages: Rocket Lab's Mission Control.

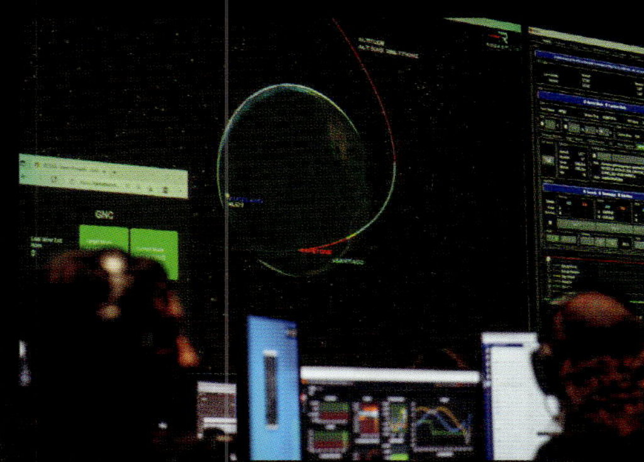

The Satellite Builder

Right: Katie Gwozdecky, Mission Manager, Rocket Lab.

Among the skilled engineers who joined Rocket Lab just after the Sinclair Interplanetary purchase was Katie Gwozdecky, a Systems Engineer and now a Mission Manager at Rocket Lab.

"We've proven our technical chops with the W-Series vehicles. Next up we're responsible for the 'ESCAPADE' Mars mission, a LEO constellation for MDA/Globalstar, and so many more unique spacecraft to service academic, commercial, and governmental causes. This is just the beginning."

— Katie Gwozdecky

Katie grew up in Canada with *National Geographic* posters of the Voyager probes and images of Saturn shot from the Cassini probe lining her bedroom walls. As an undergraduate at the University of Toronto Institute for Aerospace Studies (UTIAS), she helped start an extracurricular group that built a sustainable CubeSat development and launch program, all funded by students. It was while working nights and weekends on these CubeSats during her undergrad that she first met Doug Sinclair, a gracious mentor, who would regularly attend team design reviews to provide technical advice. Her feverish passion for satellites took her to pursue a Master's after graduation at the Space Flight Lab—a microsatellite lab associated with the University—where a number of Rocket Lab staff cut their teeth.

That led to a job at Sinclair Interplanetary that had just completed its acquisition by Rocket Lab, where she would help scale up the company's new reaction wheel production line.

"I met an amazing group of folks during the Covid-19 pandemic, and stayed to help them spool up that manufacturing capability," she said.

While Katie and others were building reaction wheels, Rocket Lab's versatile engineers were developing the Photon spacecraft. In February 2020, the company announced that both Electron and Photon would be used as part of an ambitious NASA mission to send a small satellite into lunar orbit as part of the "CAPSTONE" mission.

But it was clear to Katie that a different engineering mindset was required to enable Rocket Lab to become a big name in spacecraft design and production.

"Electron is an incredibly reliable rocket," she said. "But the choices that we make for Electron are not based so much on long-term compatibility in the space environment, but on a hours-long lifetime in orbit after launch, then decay into the atmosphere."

When it comes to spacecraft, like rockets, limiting mass is still all-important. But reliability is paramount, as most satellites can't be repaired in space. Rocket Lab's satellite engineers are able to customize Photon's capabilities and lifetime to the extent that it has now spawned three additional spacecraft formats—Explorer, Lightning, and Pioneer.

Katie soon had her own challenging Pioneer project to work on when Varda Space Industries, a microgravity-enabled life sciences company based in California, approached Rocket Lab with a plan to launch a capsule into space to conduct pharmaceutical experiments in orbit.

Varda was developing capsules capable of providing the power, communications, and connective systems required to harness microgravity for certain manufacturing processes. The unique properties of micro-gravity for example allow the creation of new pharmaceutical formulations that may reduce side effects for users, a transformational possibility. Its first missions focused on seeding the HIV drug Ritonavir in space, to determine if microgravity may ultimately produce the crystalline structure that may reduce those side effects.

To bring the pharmaceutical materials back to Earth for analysis by scientists, the W-1 capsule would have to reenter the Earth's atmosphere and survive the return journey. Pioneer needed to accompany it part of the way back, setting the capsule on course before burning up in the atmosphere.

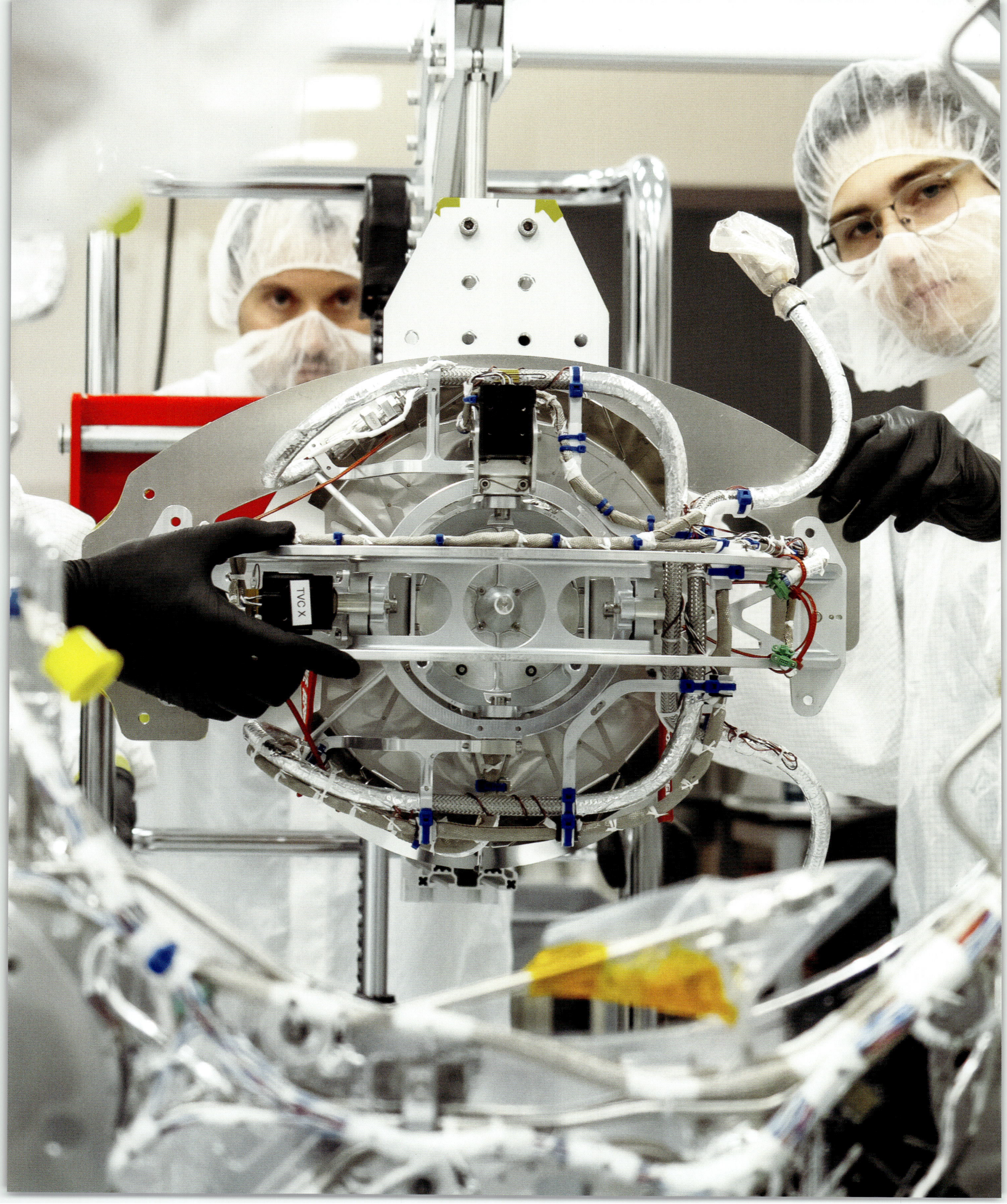
TVC X

"Two of the biggest things that the Pioneer bus brought to the table for Rocket Lab were the ability to perform targeted reentry with our epic Curie propulsion system, and rideshare launch vehicle compatibility," Katie said. "We can hit the target, every time, and we can be the safest propulsion system capable of it on the rideshare stack."

On February 21, 2024, the Varda capsule gently touched down under its parachute in the Utah desert after spending eight months integrated to a Rocket Lab Pioneer spacecraft in orbit. It was the world's first orbital manufacturing mission to take place outside the International Space Station.

The success demonstrated the potential for producing high-value pharmaceuticals in orbit. On January 14, 2025, Varda's W-2 mission was launched from Vandenberg Space Force Base with Pioneer yet again successfully carrying the manufacturing capsule into orbit before bringing it back down to Earth in March 2025. This time, Pioneer for W-2 demonstrated precision trajectory planning and landing in the Koonibba Test and Training Range in South Australia. A third mission did the same thing again just two months later in May.

Katie and the Space Systems team now have a major pipeline of satellite projects under development, including a $515-million-dollar project Rocket Lab is leading to build a constellation of eighteen satellites for the Space Development Agency (SDA). The Low Earth Orbit satellites will form a mesh network, offering more resilient communications and surveillance capability for the US military.

"We've proven our technical chops with the W-Series vehicles. Next up we're responsible for the "ESCAPADE" Mars mission, a LEO constellation for MDA/Globalstar, and so many more unique spacecraft to service academic, commercial, and governmental causes. This is just the beginning."

At the start of 2025, Rocket Lab had $720 million in satellite contract value on its books from development, to production, and operation. An increasing focus is on building, launching, and operating satellite constellations, a capability that will come into its own with the debut of Neutron.

"Even better, our work is making a real difference. Supporting Varda's reentry missions means we support pharmaceutical study that might help people's health outcomes. Supporting the MDA/Globalstar constellation means we are putting emergency communications in orbit that could make the difference in a critical moment. Supporting the SDA means we are conscious of a dynamic defense playing field," Katie said.

"I'm excited about what our department will look like in a year, let alone five or more. We may have over twenty spacecraft in orbit around Earth leveraging diverse technologies, and some on the way to other planets to conduct novel science. Glad to be part of the team making it happen."

Left: Integration of the propulsion system to one of Rocket Lab's spacecraft for the NASA "ESCAPADE" mission to Mars.

Right: Scientific instruments are integrated to Rocket Lab's twin spacecraft for the "ESCAPADE" mission.

Rocket Lab's Family of Satellites

Rocket Lab's spacecraft are designed to enable a broad range of civil, defense, and intelligence missions.

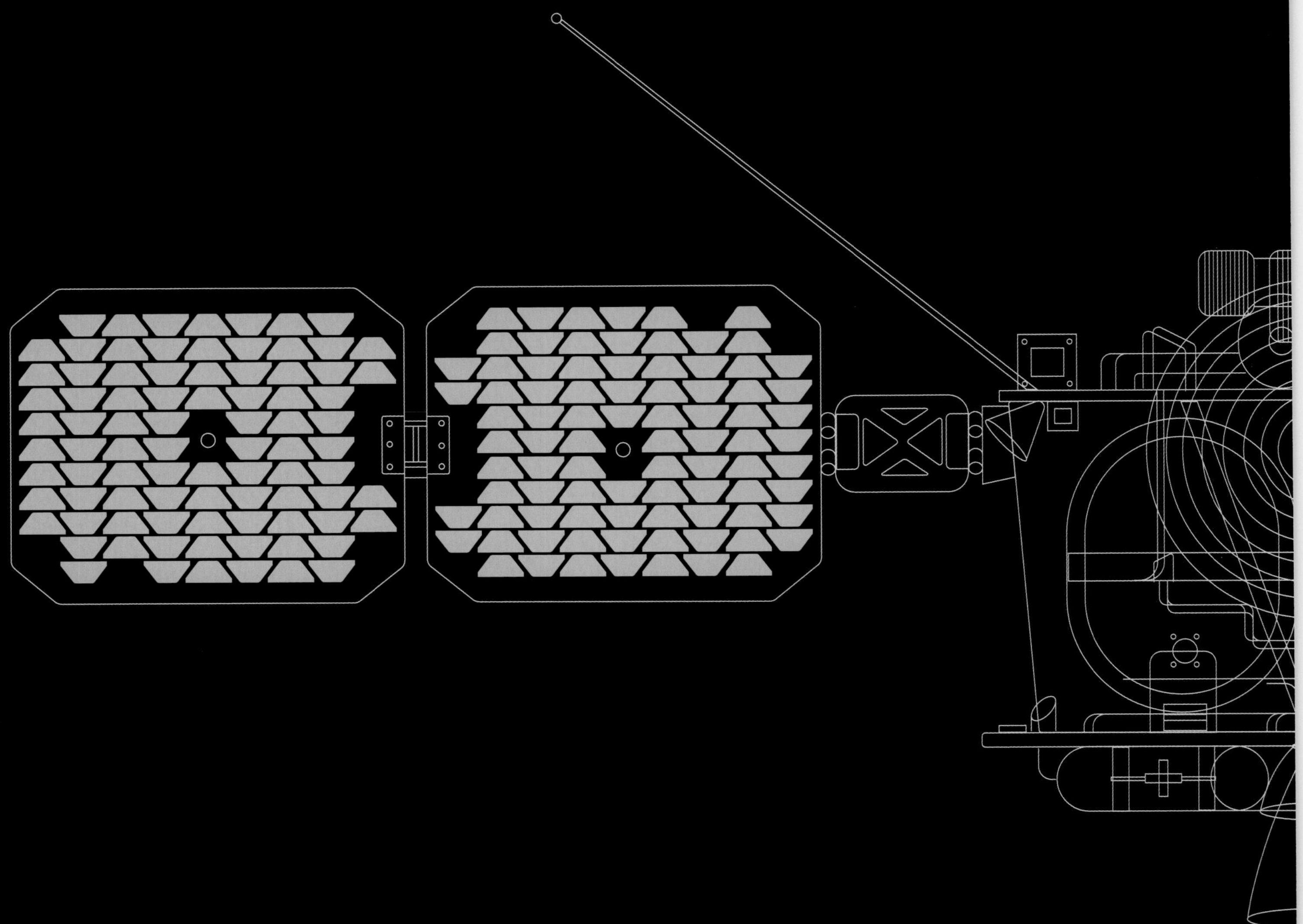

EXPLORER

Rocket Lab's interplanetary spacecraft platform.

A high delta-V spacecraft with large propellant tanks, deep space capable avionics, and ranging transponders.

Explorer enables small spacecraft missions to planetary destinations like Mars and Venus, the Moon, highly eccentric Earth orbits, geosynchronous Earth orbit (GEO), Earth-Moon Lagrange points, Earth-Sun Lagrange points, and near-Earth objects (NEOs).

Depending on the mission profile, Explorer can be launched on Electron, Neutron, or other launch vehicles. Explorer gained flight heritage during the "CAPSTONE" mission to the Moon for NASA, and forms the basis of the twin spacecraft Rocket Lab is developing for the "ESCAPADE" mission to Mars for NASA and the University of California, Berkeley.

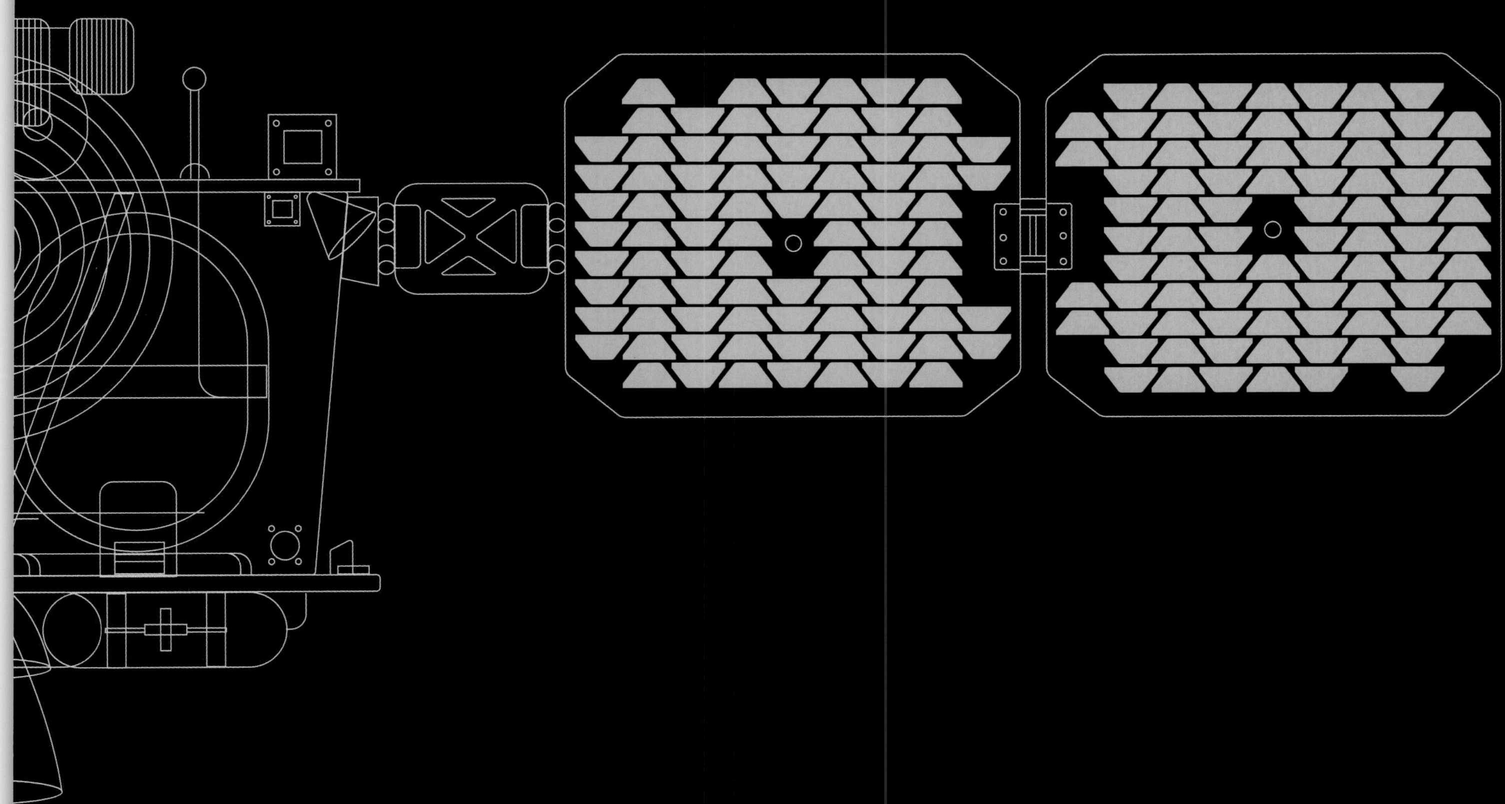

LIGHTNING

The company's newest bus for high-power long-life missions.

Rocket Lab's newest spacecraft bus is designed for a twelve-year-plus orbital lifespan in LEO, delivers high power and high radiation tolerance, incorporating redundancy in critical subsystems.

This ~3 kW bus is ideal for high operational duty cycle telecommunications and remote sensing applications. Lightning can be launched on Neutron and other medium and heavy launch vehicles. Lightning is based on the same design as the custom spacecraft Rocket Lab developed for Globalstar and is the basis for the company's bus for the Space Development Agency.

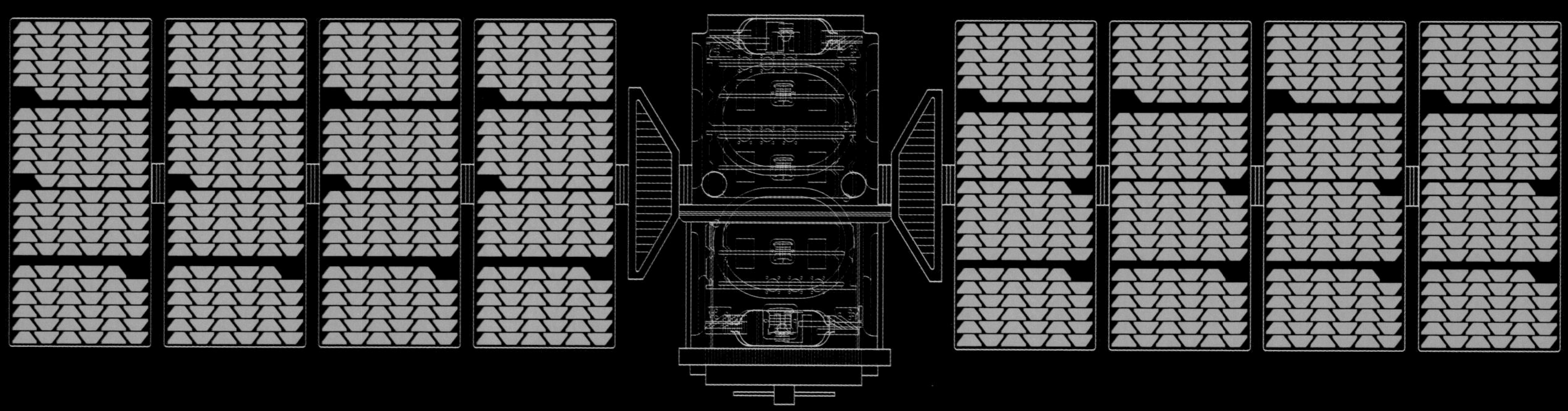

PHOTON

Rocket Lab's primary integrated launch and spacecraft solution.

The original Rocket Lab spacecraft, Photon is an integrated launch-plus-spacecraft solution. Photon is a modified Electron Kick Stage with power, propulsion, and communications to deliver a capable LEO platform that serves civil, defense, and commercial customers across a range of missions.

With a mass of 440 to 660 pounds (200 to 300 kg), Photon is launched on Electron for responsive space and other challenging missions, like cryogenic fuel demonstration. Photon takes advantage of precision orbit insertion, launch on demand, and the lowest launch environments in industry.

PIONEER

A highly maneuverable spacecraft designed to support large payloads and reentry.

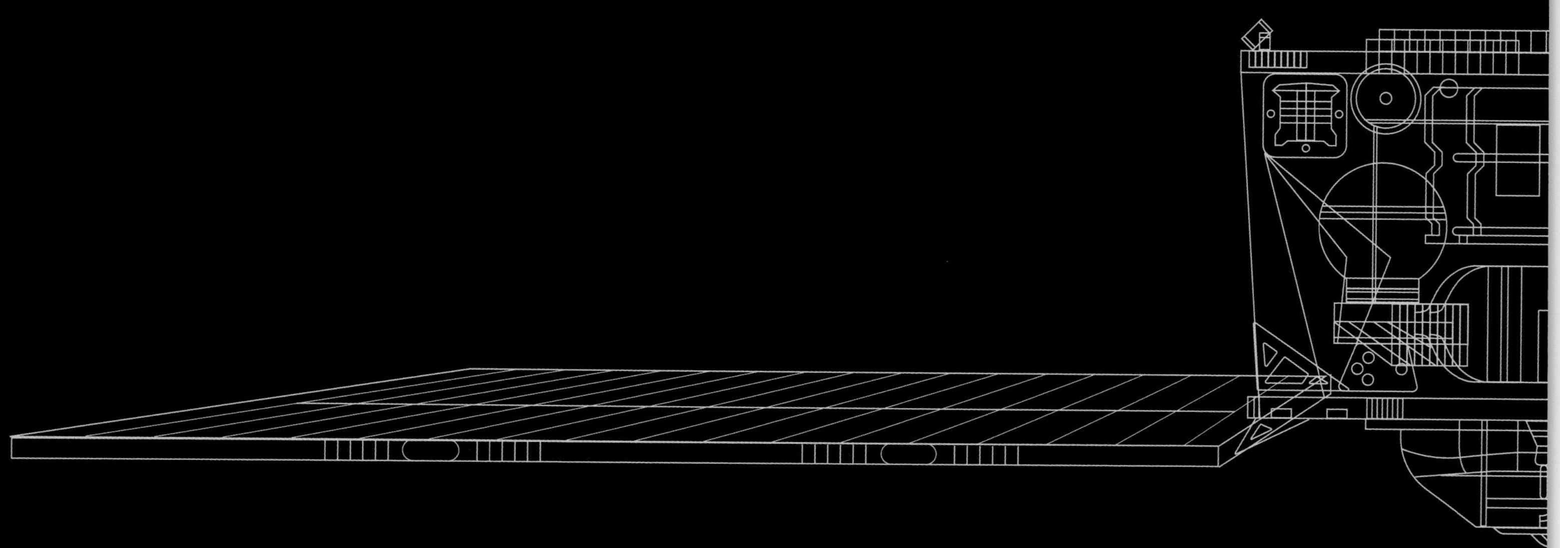

Pioneer is a medium delta-V platform that can carry payloads up to 265 pounds (120 kg) and unique mission profiles, including reentry missions for Varda Space Industries, and dynamic space operations. Depending on the mission profile, Pioneer can be launched on Electron, Neutron, or other launch vehicles.

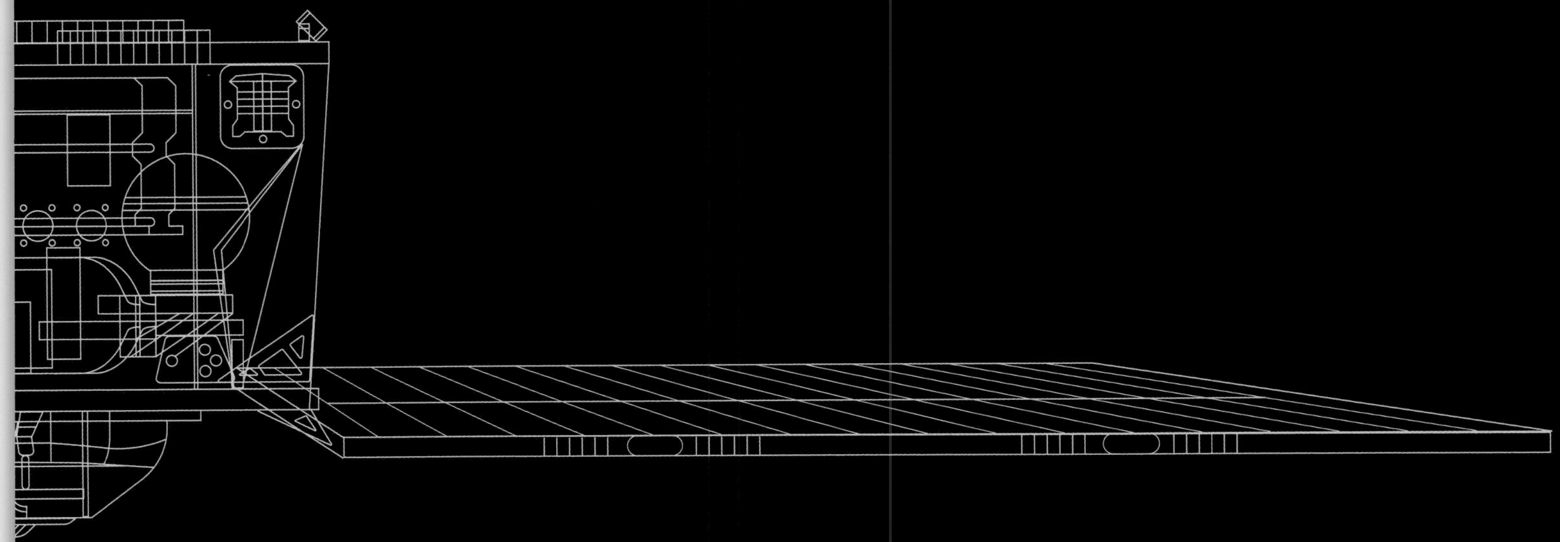

Introducing the “Flatellite”

A February 2025 Rocket Lab earnings call outlined equal measures of innovation and hustle underway at the company. It saw the debut of a novel satellite product called “Flatellite,” a flat satellite designed for mass production and customization for extensive satellite constellations. This innovative design allows for stacking during launches, potentially increasing the number of satellites deployed per mission.

“Traditional spacecraft sort of had a box stuck on the side for the satellite capability. Starlink was the first example of satellites at scale, and Flatellite is the next evolution,” said Peter. “It looks different, but the goal is the same—delivering the maximum amount of aperture per kilogram, which ultimately means the least number of launches required to build a constellation.”

That constellation could be launched on behalf of customers, as Rocket Lab has been using Electron launches to do on a smaller scale for years. But it also involves sending Rocket Lab’s own satellites into space on Neutron rockets.

“Whether it’s a communications application or Earth observation, it doesn’t matter,” said Peter. “What matters is that you have the ability to launch whenever you want to launch, and you have the ability to build whatever spacecraft you

"Flatellite is the next evolution . . . It looks different, but the goal is the same—delivering the maximum amount of aperture per kilogram, which ultimately means the least number of launches required to build a constellation."

— Sir Peter Beck

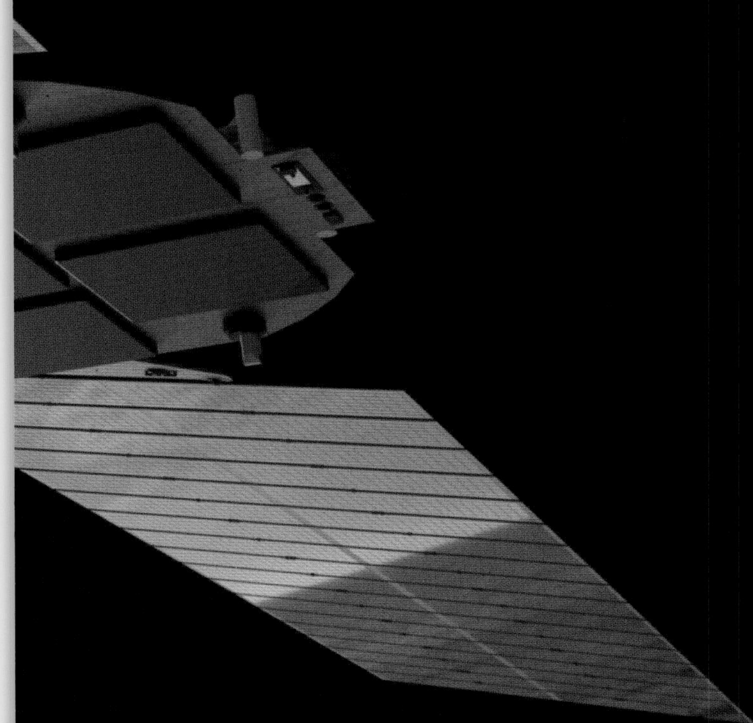

want to build. It comes back to my fundamental belief that the biggest space companies in the world are going to be companies that have the ability to deploy their own infrastructure."

The Virtues of Vertical Integration

By the start of 2025, Rocket Lab had completed the preliminary design review for a project to design, build, and operate eighteen satellites for the US Department of Defense's Proliferated Warfighter Space Architecture.

The $515-million-deal, initially announced in December 2023, cemented Rocket Lab's position as a key defense contractor. The Tranche 2 Transport Layer-Beta (T2TL-Beta) data transport satellites will form part of a cutting-edge Low Earth Orbit (LEO) constellation designed to provide secure, low-latency communications, and missile tracking capabilities for military operations.

The contract, first announced in December 2023, establishes Rocket Lab as a prime defense contractor. Under the agreement, the company will deliver satellites starting in 2027. The contract will be executed by Rocket Lab National Security (RLNS), the company's subsidiary created specifically to serve the US defense and intelligence community.

The agreement includes operating the satellites through 2030, with an option to extend operations until 2033. All eighteen satellites will feature Rocket Lab's vertically integrated subsystems and components, including solar panels, composite structures, star trackers, reaction wheels, radios, flight and ground software, avionics, and launch dispensers.

"This is the master plan, right?" Peter said. "We've been acquiring all of those satellite businesses. All of the projects that we've taken on have basically involved our customers paying us to learn all the things we need to learn to do this."

While Neutron will help Rocket Lab develop constellations from 2026 onwards, and Electron will continue to be the go-to launch vehicle for rapid turnaround, suborbital, and small satellite missions, Peter said the company is ready to step up to answer the call to explore the solar system.

Trusted With the Most Sensitive Missions

Rocket Lab received several funding lifelines in the form of contracts from the US military in its early days. Since then, it has rapidly expanded its footprint in military and national security space projects, securing a series of major contracts with the US and allied governments in 2025.

Most notably, Rocket Lab was selected by the US Space Force to compete in the National Security Space Launch (NSSL) Phase 3 Lane 1 program, a contract valued at up to 5.6 billion dollars through 2029, with at least thirty missions planned and potential extension to 2034. The company's new Neutron rocket, designed for heavier payloads and reusability, will be the launch vehicle for these high-priority national security missions.

In parallel, Rocket Lab was chosen to support the US Air Force's $46-billion-Enterprise-Wide Agile Acquisition Contract (EWAAC) and the UK Ministry of Defence's $1.3 billion Hypersonic Technologies & Capability Development Framework, providing hypersonic test launches and engineering services with its HASTE vehicle. These programs aim to accelerate the development of advanced hypersonic and reentry technologies.

"I'm passionate about national security," said Richard French, Vice President—Business Development and Strategy, Space Systems at Rocket Lab.

A triplet, Richard's sister is a commander in the US Navy and has been deployed overseas many times.

"Ensuring that the warfighter has what they need and to protect themselves and the nation has always been close to my heart," he said .

Joining Forces

Rocket Lab's Space Systems business has grown considerably through strategic acquisitions that have allowed it to become a leading supplier of infrastructure in orbit.

These acquisitions have strengthened Rocket Lab's capabilities across various aspects of the space industry, from spacecraft components and solar power products to separation systems and flight software, supporting the company's mission to pursue, as CEO Peter Beck puts it, "every part of the space value chain."

SINCLAIR INTERPLANETARY

Rocket Lab's first acquisition added flight-proven satellite components such as reaction wheels and star trackers, with heritage on missions like LightSail 2. April 2020—$17 million.

ADVANCED SOLUTIONS INC. (ASI)

Space software, mission simulation systems, guidance, navigation, and control (GNC). ASI's flight software and mission operations expertise enabled Rocket Lab to offer full mission lifecycle support. October 2021—$45.5 million dollars.

PLANETARY SYSTEMS CORPORATION (PSC):

Satellite hardware, dispensers, and separation systems brought spacecraft separation systems critical for reliable satellite deployment. November 2021—$65 to $80 million.

SOLAERO

Space solar power technology. Provided high-performance space solar power systems, dramatically improving satellite power efficiency and enabling larger payloads. January 2022—$80 million.

Rocket Lab's Space Systems—1,700 Missions and Counting

Rocket Lab's Space Systems division has evolved into a comprehensive end-to-end space solutions provider, offering a wide range of components and services for both its own missions and those of other companies.

Opening up Options for Space-based Communication

As part of a partnership with Canada-based advanced space technologies company MDA, Rocket Lab is leading the development of the spacecraft bus for Globalstar's new generation of Low Earth Orbit (LEO) satellites, which are designed to enhance mobile satellite services and direct-to-device connectivity worldwide.

Rocket Lab's contributions include providing critical satellite components such as reaction wheels, star trackers, radios, flight software, and solar power solutions. These systems ensure precise attitude control, reliable communications, and efficient power management for constellation-class satellites. Additionally, Rocket Lab's next-generation MAX Constellation flight software and InterMission ground operations platform offer advanced automation, cybersecurity, and real-time constellation management, optimizing performance and reliability for Globalstar's expanding network.

NASA James Webb
Space Telescope

The Largest Telescope in Space

Rocket Lab technology is integral to NASA's James Webb Space Telescope (JWST) through the advanced solar power hardware it provided. The telescope is powered by solar panels manufactured by SolAero Technologies, which was acquired by Rocket Lab in 2022. The panels supply the critical energy needed for JWST's instruments and systems as it orbits the Sun-Earth L2 point. These high-efficiency solar arrays ensure continuous operation of Webb's sophisticated scientific equipment, enabling its groundbreaking infrared observations and deep space imaging.

Rocket Lab's contribution demonstrates the company's expertise in space-grade power solutions, supporting one of the most ambitious and complex space observatories ever launched. The success of JWST, including capturing the sharpest images of the universe to date, highlights how Rocket Lab hardware underpins vital infrastructure for flagship NASA missions.

Previous page: A render of Rocket Lab's Lightning spacecraft for the Globalstar constellation.

Opposite: An image of the Pillars of Creation which lie within the Eagle Nebula, 6,500 light years away, taken with NASA's James Webb Space Telescope's near-infrared-light view.

This page: NASA's James Webb Space Telescope (JWST) powered by Rocket Lab technology. The company's high-efficiency triple-junction solar cells convert Sunlight to power on the JWST.

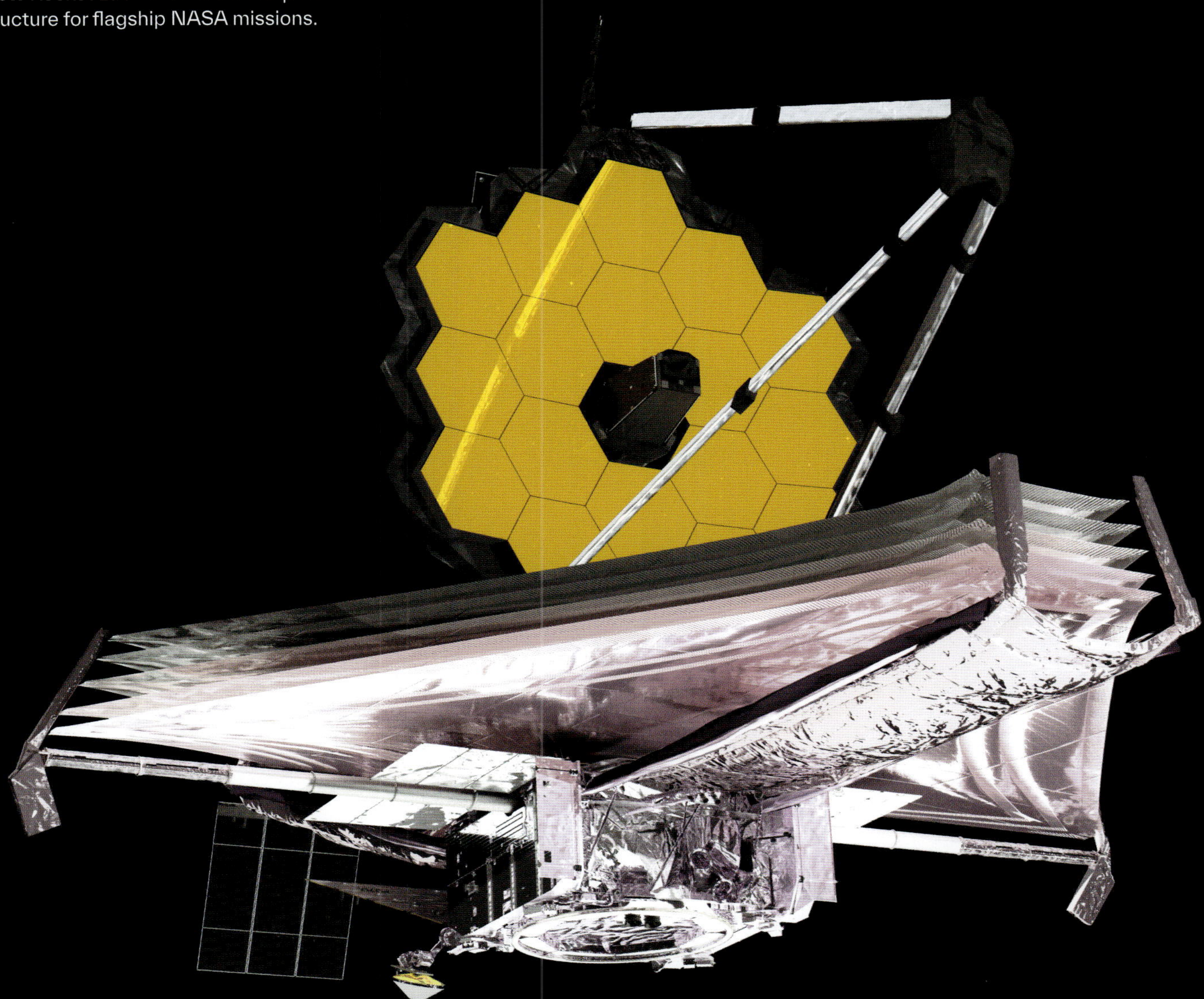

NASA Parker Solar Probe

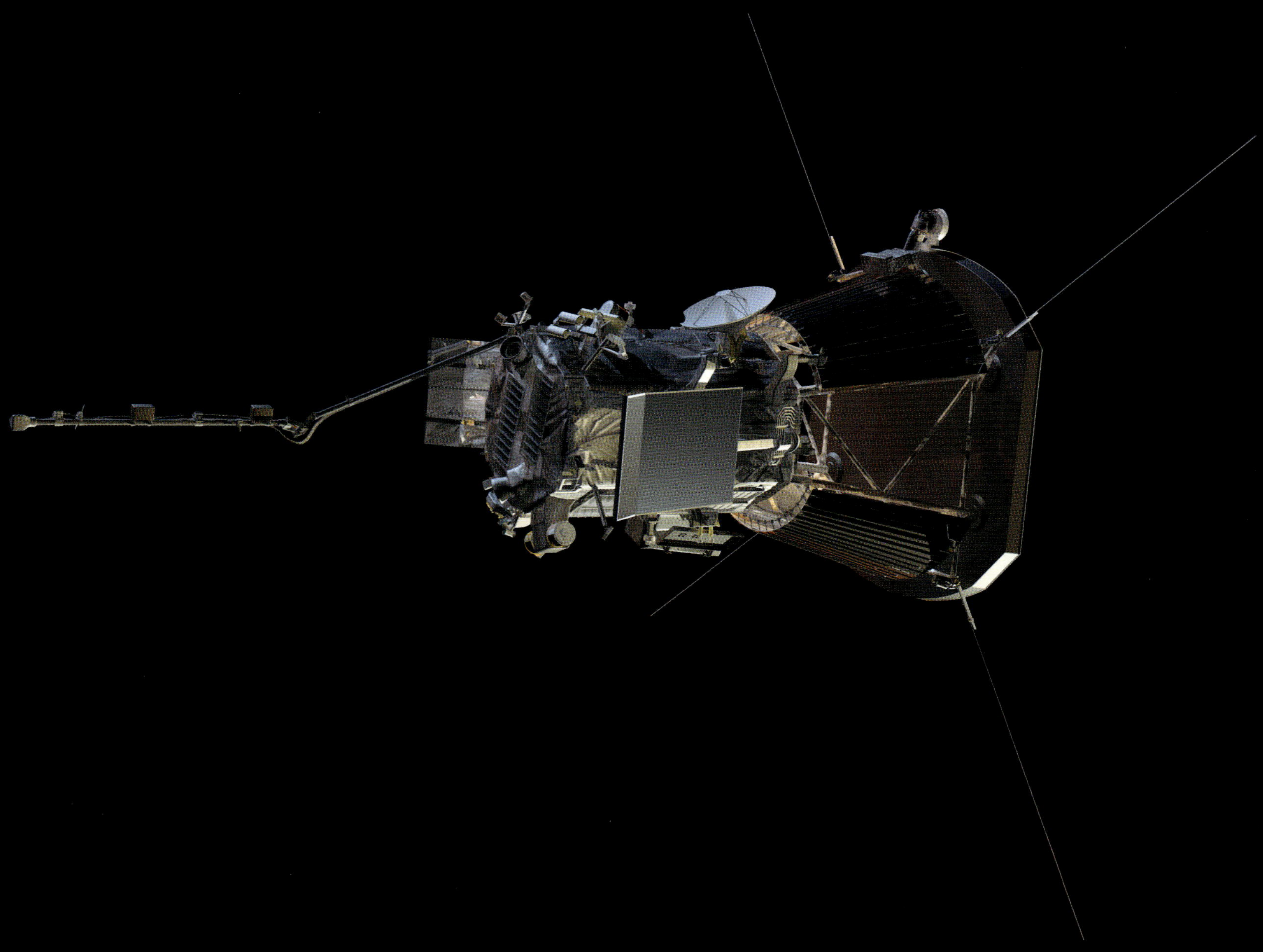

This page: NASA's Parker Solar Probe. Rocket Lab manufactured the two, first-of-its-kind, water-cooled solar panels to help keep them from overheating as the Probe closely approaches the Sun.

Overleaf: A render of NASA's Mars InSight Lander.

Energizing NASA's Mission to Touch the Sun

In 2018, NASA's Parker Solar Probe set off on a journey to the Sun, with the aim of observing the Sun's outer corona. In doing so, the spacecraft became the closest ever artificial object to the Sun.

Rocket Lab assisted the mission by supplying the advanced solar panels that power the Parker Solar Probe as it journeys closer to the Sun than any probe before it. These solar panels were developed by SolAero Technologies, a leading space solar power company acquired by Rocket Lab in 2022.

SolAero's high-efficiency, radiation-hardened solar cells are specifically engineered to withstand the intense heat and radiation near the Sun, ensuring reliable power generation even during the probe's closest approaches.

The panels are integrated with a unique cooling system that pumps fluid through channels beneath the cells, dissipating the extreme heat and maintaining optimal operating temperatures. Rocket Lab's contribution has enabled the Parker Solar Probe to conduct groundbreaking science, collecting unprecedented data on the Sun's corona and solar wind, and revolutionizing our understanding of solar phenomena. The spacecraft entered its final orbit on November 6, 2024 and was still active and returning observation data to Earth as of mid-2025.

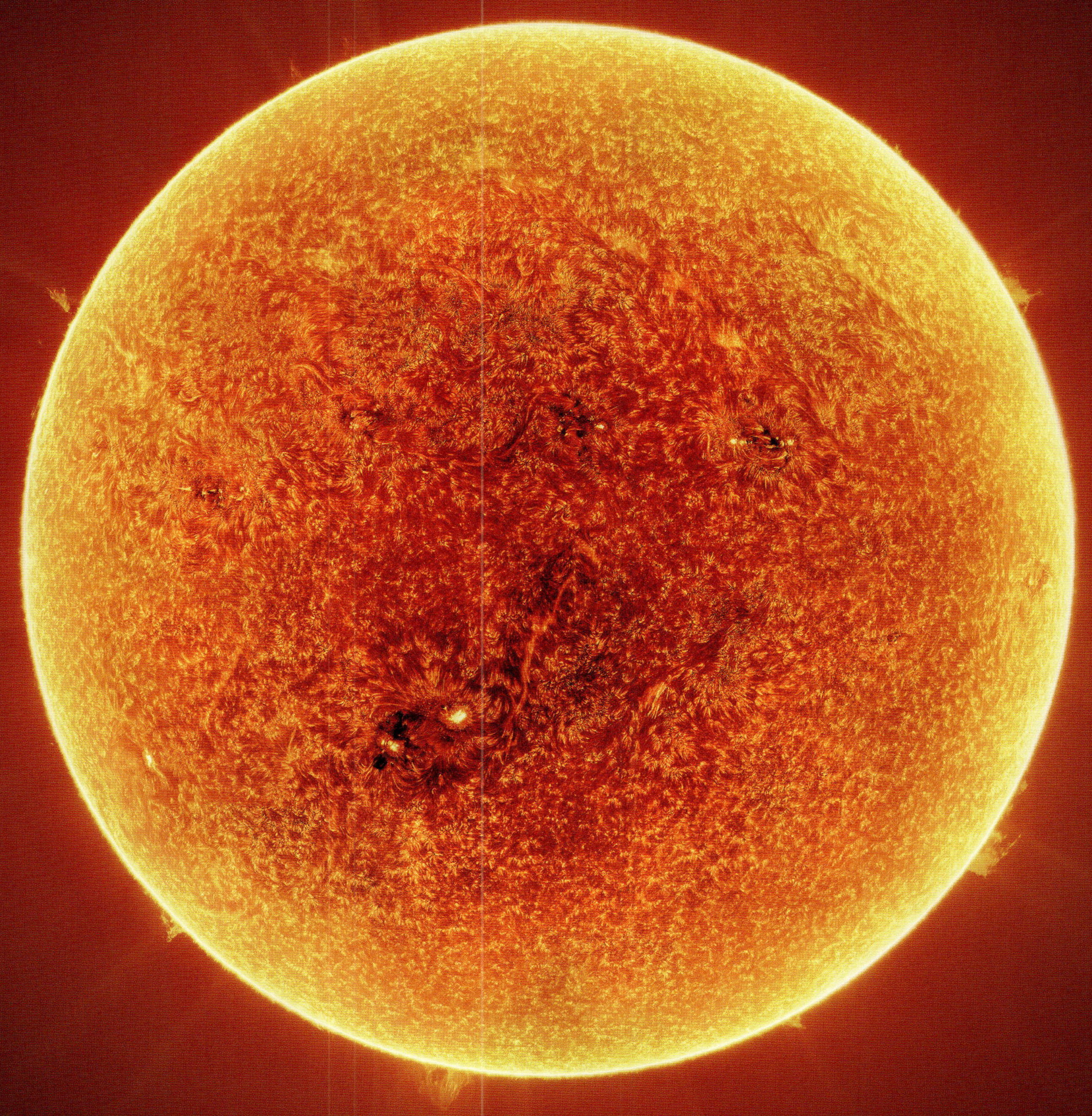

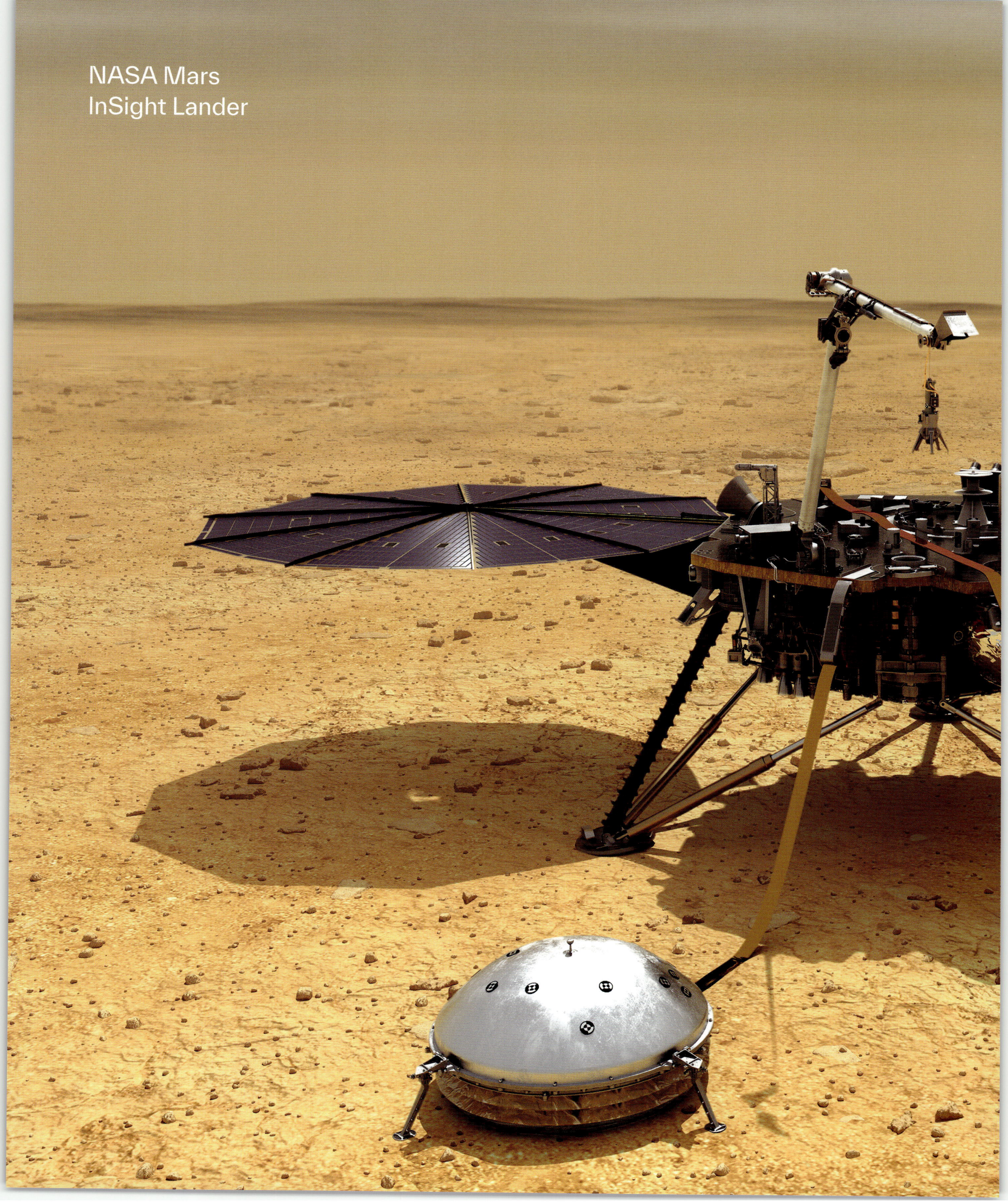
NASA Mars
InSight Lander

Lighting the Way for InSight on Mars

The Mars InSight lander was a stationary NASA mission designed to give the Red Planet its first thorough internal examination. Unlike previous Mars rovers that explored the surface, InSight's primary goal was to study the deep interior of Mars to uncover how rocky planets like Earth, Venus, and Mars formed and evolved over billions of years.

Launched on May 5, 2018 from Vandenberg Space Force Base in California, InSight successfully touched down on the Red Planet Elysium Planitia on November 26, 2018. The lander operated on the surface of Mars for over four years, until 2022 when its solar panels gathered too much dust to operate. Those solar panels were made by SolAero and provided all the energy for the lander's scientific instruments and daily operations on the Martian surface.

The high-efficiency solar arrays were the largest ever deployed on Mars at the time, enabling InSight to conduct groundbreaking research into the planet's interior.

NASA Ingenuity Mars Helicopter

This page and opposite: A render of NASA's Ingenuity Mars Helicopter.

A Helicopter Ride on Mars

Rocket Lab played a crucial role in NASA's Ingenuity Mars Helicopter mission by supplying the solar panels that powered the helicopter throughout its historic flights on Mars.

Ingenuity, which arrived on the Red Planet in 2021 attached to the Perseverance rover, relied on a custom-built solar array to recharge its lithium-ion batteries, enabling up to 90 seconds of flight per Martian day in Mars' thin atmosphere.

Rocket Lab's contribution was vital, as the solar panels provided the energy necessary for Ingenuity's avionics, sensors, and flight motors to function in the harsh Martian environment. The success of Ingenuity, which completed 72 flights over nearly three years—far surpassing its original five-flight goal—demonstrated the reliability and advanced engineering of Rocket Lab's hardware in supporting groundbreaking extraterrestrial aviation and NASA's broader exploration objectives.

CAPS

Rocket Lab's own Moonshot

TONE

One of the most complex, ambitious, and drama-filled projects Rocket Lab has undertaken was the "CAPSTONE" mission, which pushed people and technology to their limits but was ultimately a triumph of innovation and execution.

In 2017, NASA formally revealed plans for Artemis, a follow-up program to the famed Apollo missions of the 1960s, which aimed to land humans back on the Moon in the 2020s.

The centerpiece of the plan was the Gateway space station, that would operate in orbit around the Moon as a place for astronauts to live and work between trips to the lunar surface.

Like the International Space Station, Gateway would serve as a habitation module, scientific lab, and communications hub. The US was finally going back to the Moon, but extensive preparations for Gateway were needed to identify an appropriate lunar orbit for the first space station that would ever orbit the Moon.

The "CAPSTONE" (Cislunar Autonomous Positioning System Technology Operations and Navigation Experiment) mission's primary objective was to test and verify the calculated orbital stability of the Near Rectilinear Halo Orbit (NRHO) around the Moon—the orbit planned for Gateway.

The Halo Effect

It was calculated to give Gateway and spacecraft docking with it the best combination of fuel efficiency and a pathway to the lunar surface.

As NASA explained: "Hanging almost like a necklace from the Moon, NRHO is a one-week orbit that is balanced between the Earth's and Moon's gravity. This orbit will periodically bring Gateway close enough to the lunar surface to provide simple access to the Moon's South Pole where astronauts will test capabilities for living on other planetary bodies, including Mars."

The orbit also gives astronauts access to other landing sites around the Moon.

A CubeSat would need to be launched into lunar orbit to gather data that would help NASA decide exactly how to position Gateway.

Peter Beck had always been fascinated with interplanetary spaceflight. But Rocket Lab had to date been involved in putting spacecraft into orbit 248 to 341 miles (400 to 550 km) above the Earth. Carrying a satellite beyond Earth's orbit all the way to the Moon, over 238,606 miles (384,000 km) away, was an entirely different job, and not one Peter was confident Electron could pull off.

He had earlier put Master's student and Rocket Lab intern Richard Hunter to work on a thought experiment to figure out how to use Electron to get a payload into orbit around Venus, a planet ripe for further scientific examination. When NASA went looking for partners for "CAPSTONE," Peter asked him to run the numbers on a lunar orbit.

"It was literally applying some really simplified equations of motion to figure out how feasible it would be to modify Electron's Kick Stage, launch it on an Electron rocket, and get something, albeit pretty small, to the Moon," Richard remembered.

Beyond Earth's Orbit

The initial feasibility study Richard conducted was enough to convince NASA to hand the job to Rocket Lab, which by that point had put numerous satellites in orbit for the agency.

"I got a call from NASA headquarters at 3 a.m. and jumped up, super excited, punching the sky for the first thirty seconds," said Peter. "Then, for like the next two and a half years, it was sheer pain and misery."

From the beginning, the mission was ambitious. It required the design and development of a new orbital engine (HyperCurie) and spacecraft (Lunar Photon) to propel the 81 pound (37 kg) NASA payload into a translunar trajectory. However, as the program matured through detailed design, system complexity grew, mass increased, and performance margins became negative, and significant modifications to Electron were required to make up the shortfall.

"Ultimately, the rocket that took off the pad lifted 705 pounds (320 kg)," said Peter. "There was no camera onboard, because we couldn't spare seven ounces (200 g). There was a three-hour argument about whether we could afford the mass of the NASA sticker on the side of it."

"As the project developed, we realized the system was more complicated than expected, which meant more weight. We needed to make the rocket more powerful, the batteries more efficient," said Richard.

Rocket Lab's experts combed through Electron's components looking for ways to shed weight and extract more power. As they proposed tweaks, Richard fed them into computer simulations, running hundreds of different configurations to see which would get Electron and its precious cargo into a lunar orbit.

The second-stage tanks were stretched, batteries upgraded, and countless small optimizations aimed at reducing weight.

Rocket Lab was initially given a tight twelve-month deadline to prepare for "CAPSTONE." But it became clear that the various modifications required would push out the timeframe. The Covid-19 pandemic also put pressure on Rocket Lab's supply chain, leading to delays.

Desirée Brundin joined Rocket Lab as Team Lead—Space Systems AIT & Systems Engineering in August 2021 and was immediately assigned to work on the Lunar Photon for "CAPSTONE." She didn't know at that stage what she had let herself in for.

"I came here expecting everything to be done. I was worried that I would miss all the fun," she said. "But there was no spacecraft. I was like, where is it?"

"As the project developed, we realized the system was more complicated than expected, which meant more weight. We needed to make the rocket more powerful, the batteries more efficient."

— Richard Hunter, Manager—Mission Design & Operations, Rocket Lab

Previous page: Rocket Lab's Lunar Photon with NASA's "CAPSTONE" spacecraft.

This page: The Electron rocket on the launch pad that launched the "CAPSTONE" mission to the Moon.

"I got a call from NASA headquarters at 3 a.m. and jumped up, super excited, punching the sky for the first thirty seconds, then, for like the next two and a half years, it was sheer pain and misery."

— Sir Peter Beck

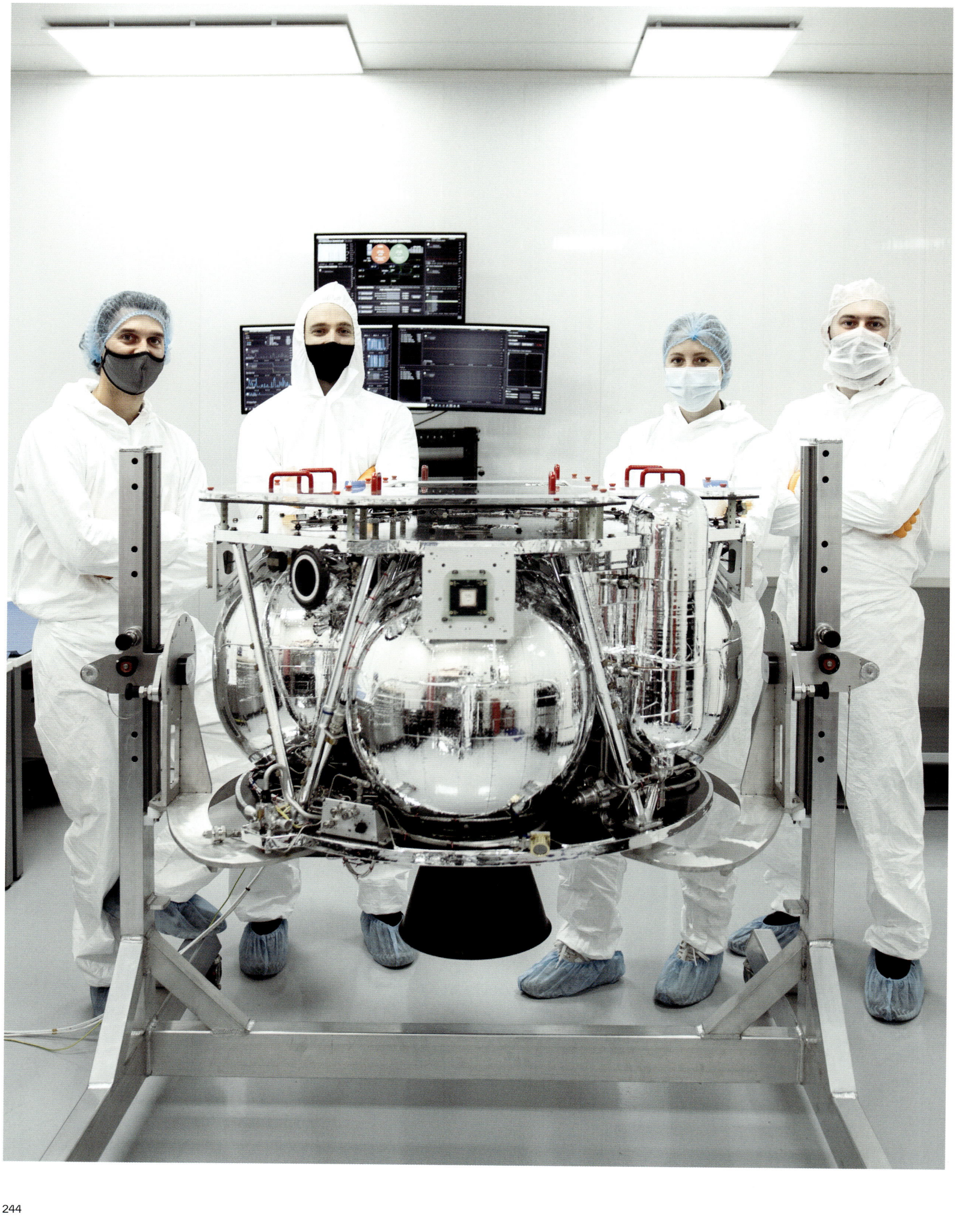

Mission Creep

Delays in delivery of key parts and technical difficulties meant that Desirée Brundin and her colleagues ended up undertaking assembly integration and testing of the Lunar Photon, and the launch campaign and mission operations, in less than a twelve-month period.

Previous page and these pages: Spacecraft integration and assembly of Rocket Lab's Lunar Photon spacecraft for the "CAPSTONE" mission.

"It's just crazy what we accomplished in so little time," she added.

One of the key challenges the team faced was developing the HyperCurie engine, a major technological advancement over the original Curie engine Rocket Lab used to maneuver the Photon spacecraft.

Unlike the pressure-fed Curie engine, the HyperCurie uses electric pumps to deliver propellant to the combustion chamber. This design choice eliminated the need for complex turbomachinery typically found in gas generator cycle engines, while allowing for more efficient operation in deep space.

A typical Electron launch is complete within an hour from lift-off to payload deployment. For the "CAPSTONE" mission, even after the initial launch to orbit on Electron was completed, Rocket Lab still had to operate the Photon spacecraft for six days in challenging conditions.

Unlike regular Electron launches to Low Earth Orbit, which are automated once in flight, Richard Hunter knew that Mission Control would have to make adjustments to Photon's path throughout the seven-day mission to the Moon.

"With that level of complexity, you just couldn't model it. We knew we would have to solve problems on the fly and send commands to the spacecraft in real time."

Left: Desirée Brundin, Team Lead—Space Systems AIT & Systems Engineering, Rocket Lab.

"With that level of complexity, you just couldn't model it. We knew we would have to solve problems on the fly and send commands to the spacecraft in real time."

— Richard Hunter

This page: Vehicle integration and assembly of the "CAPSTONE" mission.

Opposite: Lift-off of Rocket Lab's "CAPSTONE" Electron launch.

Overleaf: A rocket trajectory at night.

NASA

The Super-Maneuver

The Lunar Photon then performed a series of orbit-raising burns over six days, gradually increasing "CAPSTONE's" velocity. The plan was to perform eight burns, strategically timed to maintain sufficient momentum in space. But mid-mission, a worrying call came from Mission Control.

"In a lot of respects, it just seemed impossible. We always challenge ourselves here, but 'CAPSTONE' seemed like a step too far. Like we have in the past, we pulled it off."

— Shaun O'Donnell, Chief Engineer—Special Projects, Rocket Lab

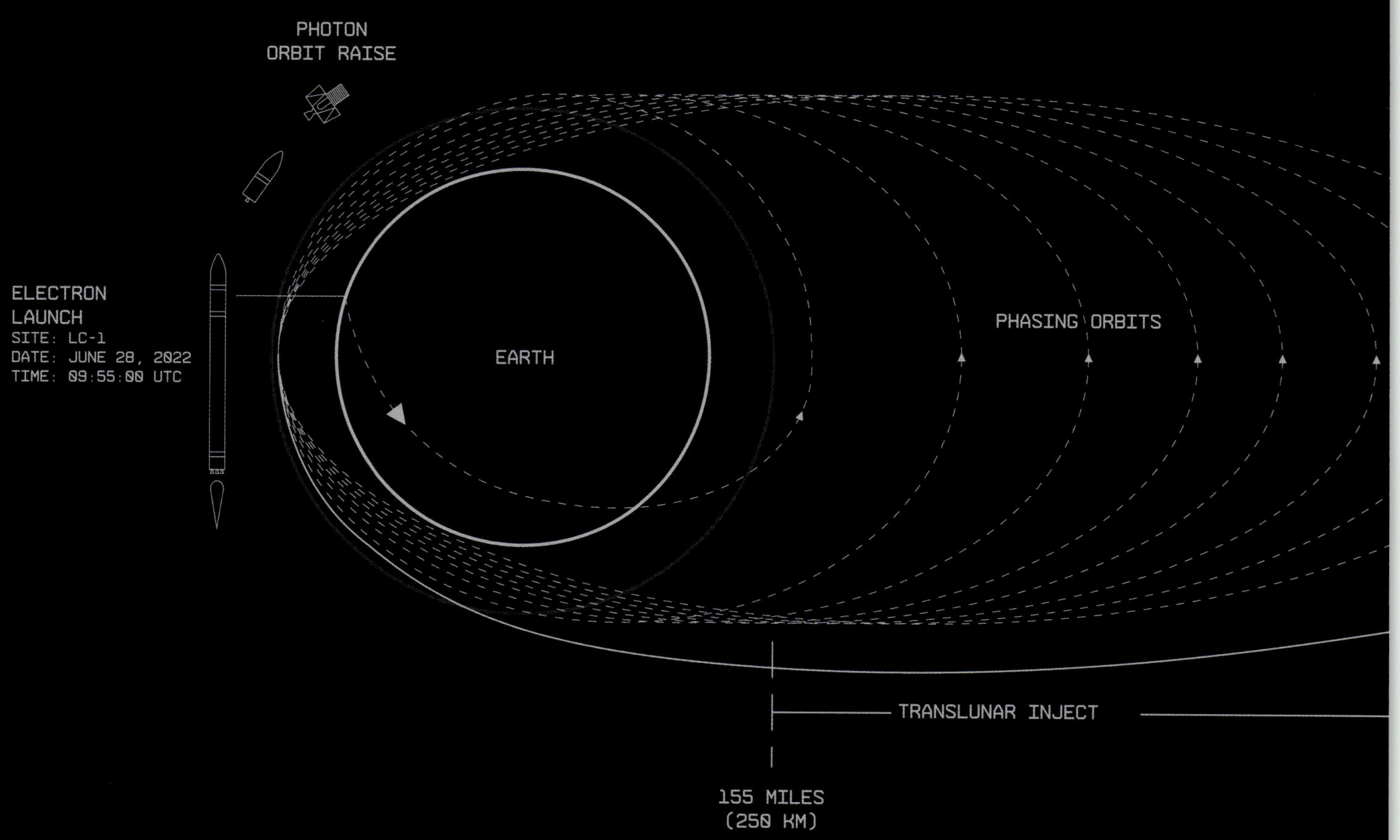

Richard was home, about to go to bed after a 20-hour stint in Mission Control, when the phone rang. "The propulsion team had been busy analyzing telemetry from the spacecraft and spotted a problem," he said. "We needed to limit the number of maneuvers we were doing. They thought it would be less risky to combine the next two burns into one and limit the number of engine restarts. This required an off-script redesign of the mission sequence," he explained.

The problem was that with each burn maneuver, the Lunar Photon's propulsion system, powered by the HyperCurie engines, was getting hotter, the result of a chemical reaction in the fuel system coking up the engine system.

"The next burn we did, it went five degrees above the thermal margin. And with the next burn it went five degrees above the margin again," said Peter. "We had eight burns to do. The engine was going to explode on the last burn to the Moon. So you know, that wasn't going to work."

The team again worked through the night to come up with a new plan dubbed the "super maneuver." Rather than eight burn maneuvers, the team would initiate seven to avoid damaging the engines, with one long burn considered to be much safer. It was risky—the team hadn't modeled that particular scenario. The longer the burn, the bigger the potential trajectory error, which could affect tracking and trajectory accuracy.

But by the time Richard got to Mission Control the next morning, the whole new maneuver had been designed and was ready to upload to the Lunar Photon.

"I sat down in my chair and watched this 3,280-feet-per-second [thousand-meter] burn go off. I was mentally prepared to run an emergency search pattern to track down and recover a lost spacecraft from non-deterministic orbit. I was like, "How did we get here in the space of one night?'"

But the improvisation paid off. When telemetry lit up after the super-maneuver, it confirmed that propulsion and GNC systems had executed flawlessly, and Lunar Photon remained on course.

On July 4, 2022, Lunar Photon successfully completed the trans-lunar injection maneuver, a critical step that took "CAPSTONE" out of Earth's orbit and on a transfer trajectory to the Moon. Another important mission milestone was marked.

"We did that trans-lunar injection burn, and you know it was flawless," said Peter. "Moreover, we had 13 percent propellant left in that vehicle."

"CAPSTONE" remains one of the proudest moments for Rocket Lab.

"In a lot of respects, it just seemed impossible," said Shaun O'Donnell, who had joined Peter all those years before with the simple goal of building and launching a small suborbital rocket. "We always challenge ourselves here, but "CAPSTONE" seemed like a step too far. Like we have in the past, we pulled it off."

"That was a tough mission," admitted now Deputy Chief Engineer Lachlan Matchett. "We had to really think out of the box, make everything as lightweight as we could, as high-performance as we could. We did an awesome job," he said.

For Morgan Connaughton, who Peter tasked with livestreaming the final critical moments of the mission, it was both exhilarating and terrifying to watch. "At that point, I was so used to a launch lifting off and you'd be in orbit nine minutes later. An hour later the payload is deployed. This was six days of terror, because all of a sudden we've done the launch, but now we are operating the spacecraft."

"We were very late on that mission, and we did that at a firm fixed price," said Peter. "So we lost money on that mission. That was not a good economic mission for us. But that mission enabled us to win the Mars "ESCAPADE" missions from NASA."

The experience was a baptism of fire for Desirée, but left her craving more. "I'm excited to do it all again. I think we should go to Venus and Mars," she said. "Just launching with your own rocket is such a special experience, and you get so much flexibility. Not a lot of companies get to do that."

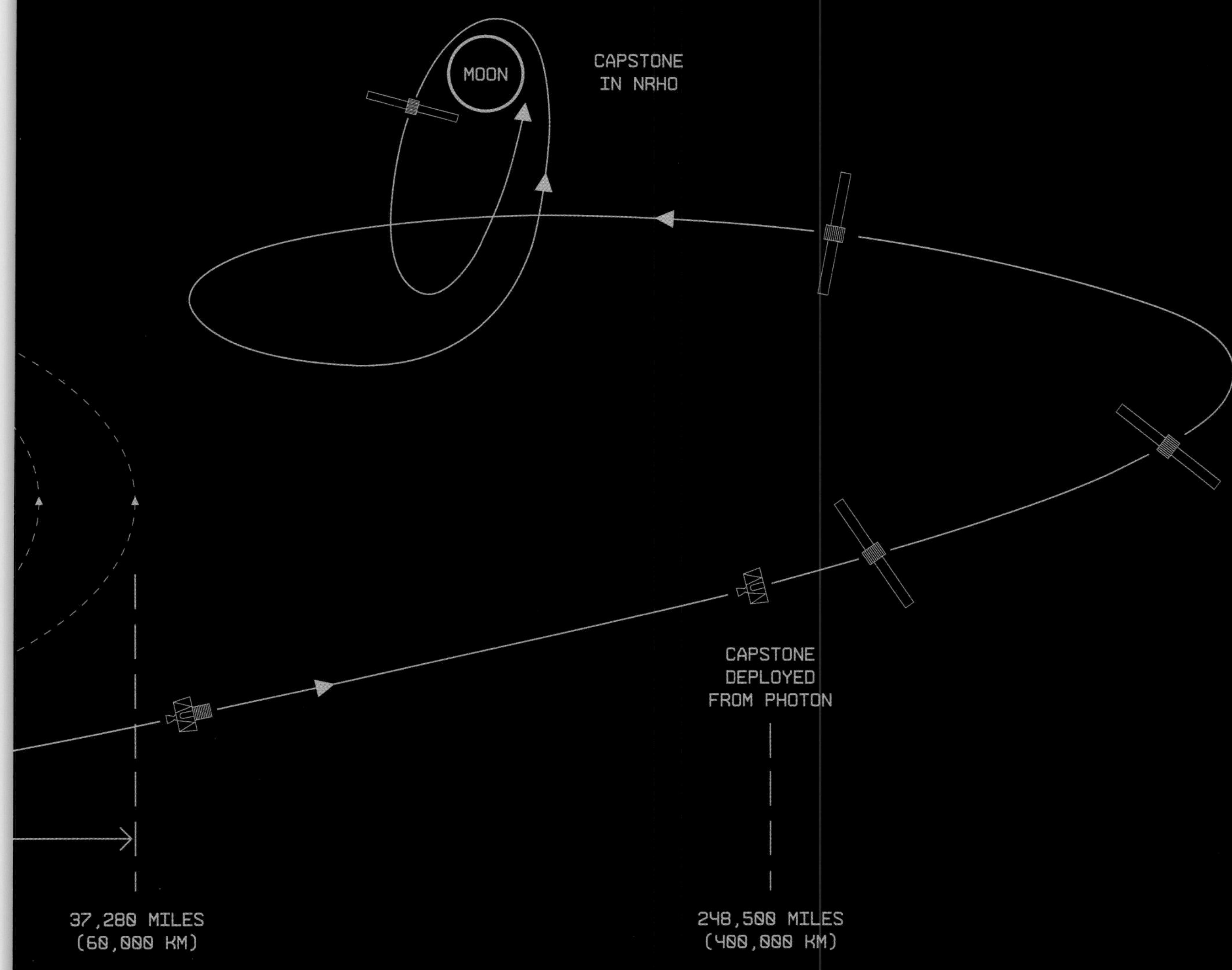

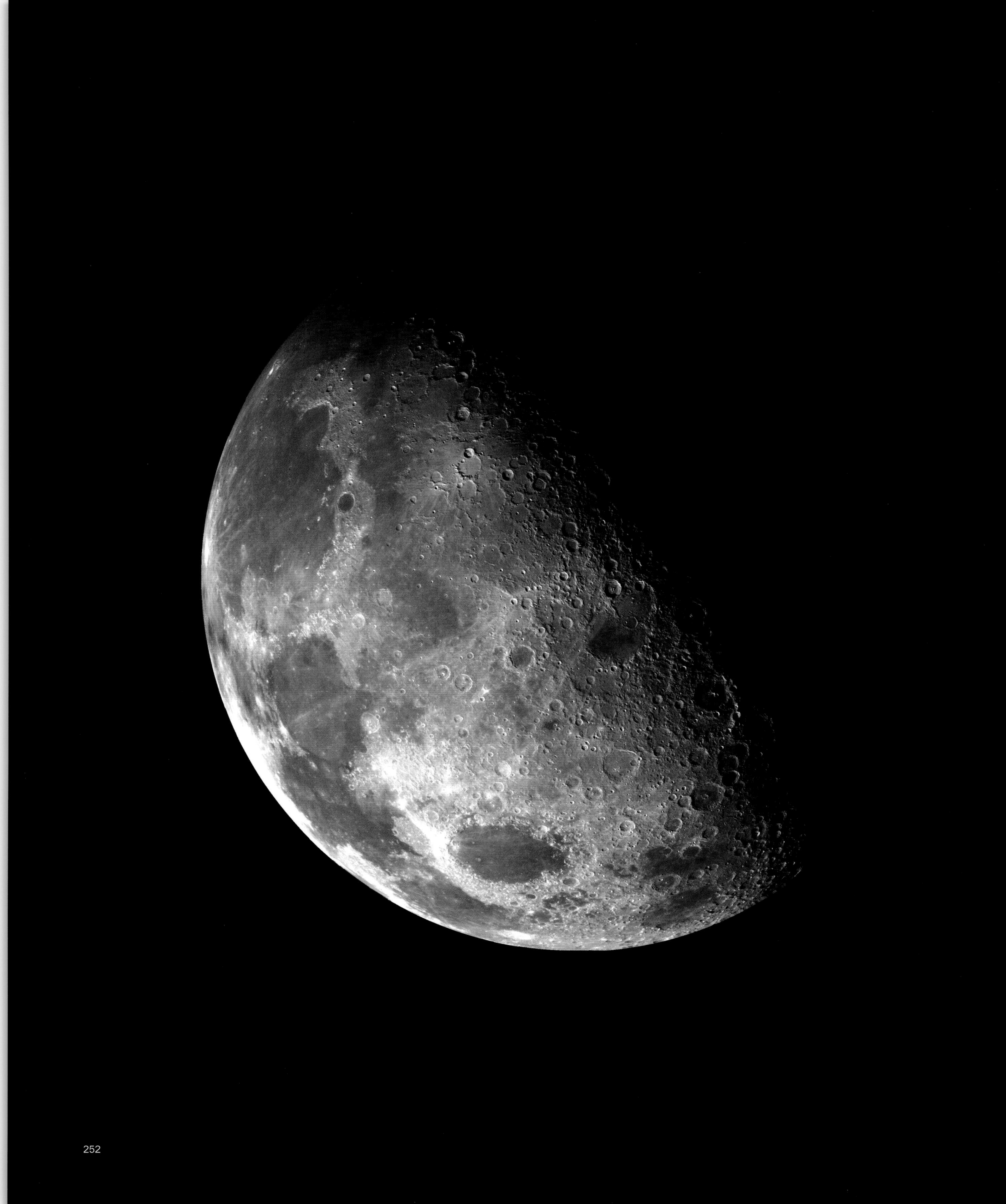

"CAPSTONE" Moon Mission Navigational Charts

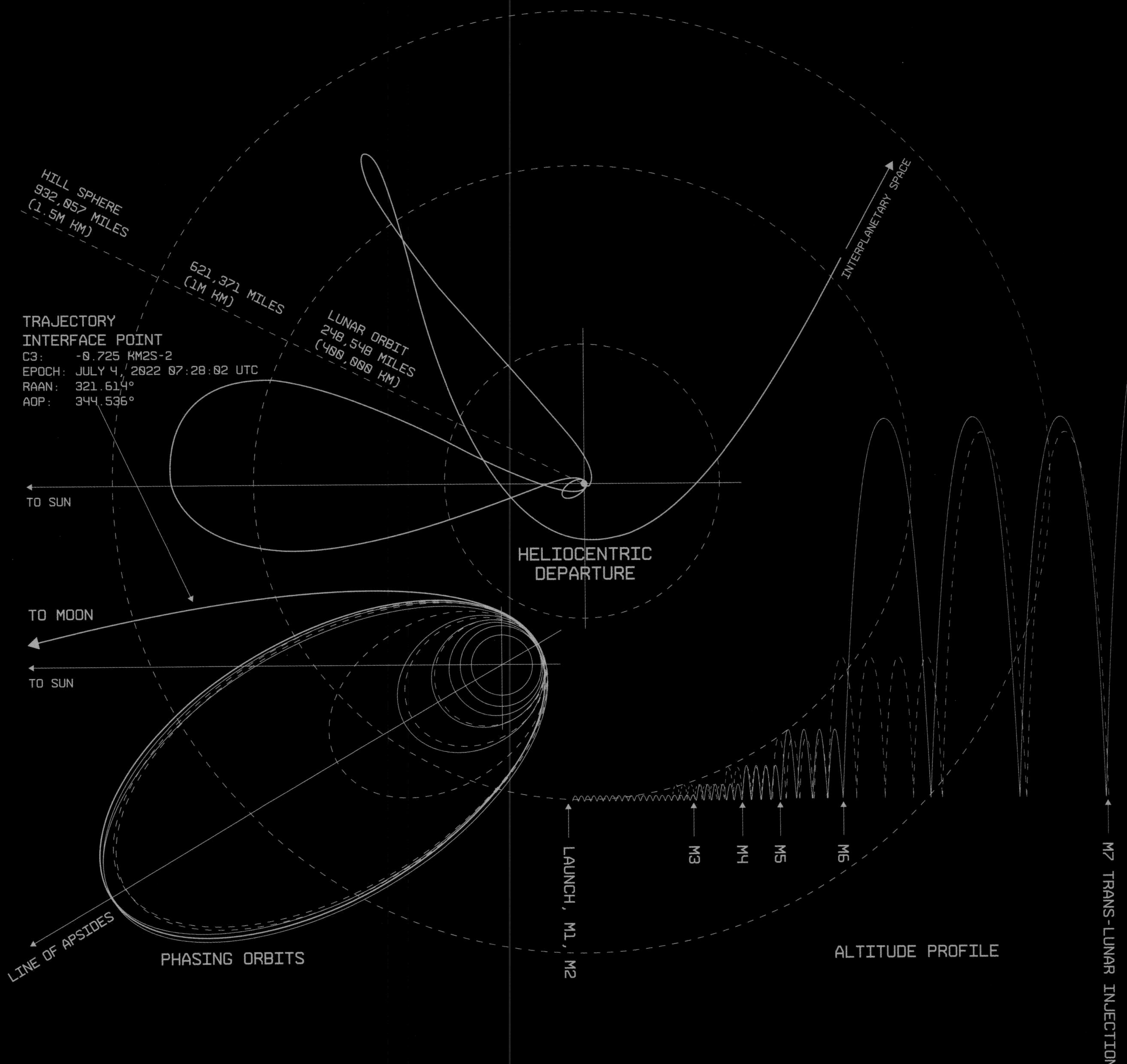

Beyond Low Earth Orbit—The HyperCurie Engine

Rocket Lab's HyperCurie engine represents a significant advancement in spacecraft propulsion technology, designed specifically for the company's Photon satellite bus.

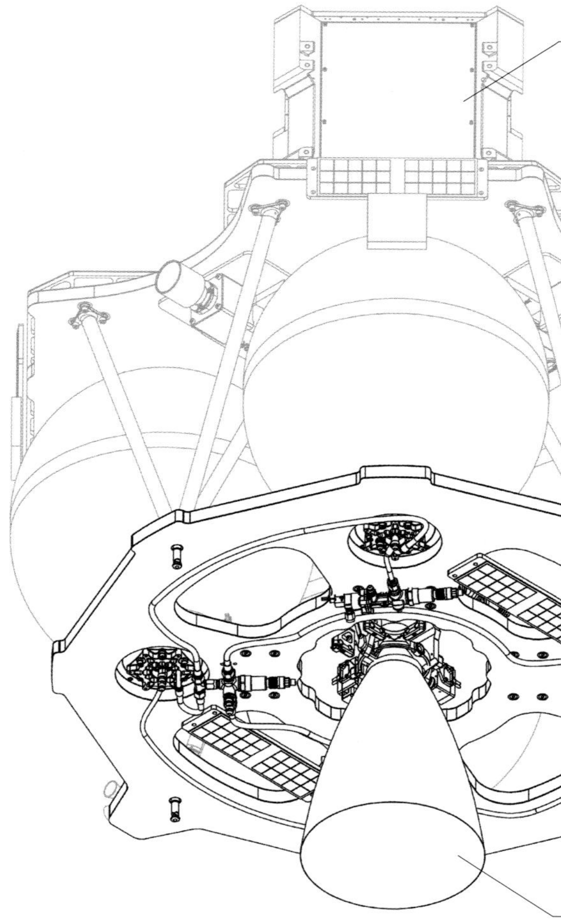

This engine is an evolution of the original Curie engine, which powers Rocket Lab's Kick Stage for the Electron rocket. The HyperCurie was developed to enable deep space missions beyond Low Earth Orbit, allowing Photon to serve as a versatile platform for lunar and interplanetary missions.

Unlike the pressure-fed Curie engine, HyperCurie is electric pump-fed, which allows it to use solar cells to recharge its batteries between burns. This innovative approach provides greater efficiency for long-duration missions.

The engine uses hypergolic propellants with electric pumps and larger propellant tanks to produce higher thrust. While the specific propellants remain undisclosed, this configuration enables the HyperCurie to deliver over 1.8 miles per second (3 km/sec) of delta-V, providing Photon with the capability to escape Earth's gravity and reach distant destinations.

HyperCurie made its debut on NASA's "CAPSTONE" mission in 2022, where it performed a series of orbit-raising maneuvers before accelerating the spacecraft to 24,500 mph (39,400 km/h) to set it on a ballistic lunar transfer trajectory.

The development of HyperCurie aligns with Rocket Lab's strategy to expand beyond launch services into spacecraft manufacturing and mission services. By creating a high-energy propulsion system for Photon, Rocket Lab has enabled low-cost access to lunar and planetary destinations for small payloads.

The company has ambitious plans for HyperCurie-powered Photon missions, including upcoming ventures to Mars and Venus. Like other Rocket Lab engines, HyperCurie is manufactured using 3D-printing techniques, allowing for rapid production and iteration.

KEY MI

Every mission Rocket Lab is involved in is unique and special in its own way. By June 2025, 232 satellites had been sent into orbit on Electron rockets, cementing Rocket Lab's position as the world's leading small launch vehicle provider.

SSIONS

Beyond launches, Rocket Lab's Space Systems components have been integral to over 1,700 missions globally, from the separation system for HawkEye 360's radio frequency-monitoring constellation, to the vertical integration of subsystems (solar panels, avionics) for eighteen satellites in the US Space Development Agency's Tranche 2 Transport Layer-Beta (T2TL-Beta) program.

But some missions are more novel, ambitious, or special from a scientific or engineering standpoint. Some missions made history, and a strong slate of upcoming missions could add new entries to the record books.

VARDA
SPACE INDUSTRIES

W-SERIES

2024, 2025

"We had to take what we'd learned from 'CAPSTONE,' and turn it on its head. Rather than flying to the Moon, we were flying back to Earth."

— Richard Hunter

These pages and overleaf: Assembly and integration of Rocket Lab's Pioneer spacecraft with integrated Varda manufacturing capsule for the first W-1 mission.

The W-1 mission undertaken by Varda Space Industries was the first fully autonomous pharmaceutical production experiment conducted outside the International Space Station.

Varda's manufacturing reactor successfully synthesized ritonavir crystals (an HIV antiviral drug) in microgravity, demonstrating improved crystal structures that could enhance drug efficacy and reduce side effects.

Varda's capsule also doubled as a hypersonic testbed, with thermal protection systems developed in partnership with NASA's Ames Research Center. Rocket Lab's Pioneer spacecraft were adapted for the missions, the company's first requiring reentry of a satellite through Earth's atmosphere so that their precious cargo could be recovered.

"We had to take what we'd learned from "CAPSTONE," and turn it on its head," said Rocket Lab's Richard Hunter. "Rather than flying to the Moon, we were flying back to Earth. We had to design a maneuver sequence to optimize the flight path angle and velocity, and be as accurate as possible."

SpaceX launched W-1 on Transporter-8 in June 2023. Rocket Lab's Pioneer spacecraft enabled precision deorbiting and reentry of Varda's capsule. When the capsule touched down in the Utah desert in February 2024 with a landing accuracy of <0.05 percent, it marked the first commercial spacecraft reentry and recovery on US soil.

The eighth-month mission showcased Rocket Lab's ability to provide power, communications, and attitude control during extended orbital operations. Varda's W-2 capsule was launched on January 14, 2025, on a Transporter-12 rideshare mission with SpaceX from Vandenberg Space Force Base in California, and successfully returned to Earth on February 27, 2025, landing at the Koonibba Test Range in South Australia. W-2 spent approximately six weeks in orbit before reentry. This mission carried an expanded pharmaceutical reactor, along with a spectrometer from the Air Force Research Laboratory.

The third capsule (W-3), also based on Rocket Lab's Pioneer spacecraft platform, was successfully deployed to orbit on March 15, 2025 and was successfully returned to Earth on May 14—completing a groundbreaking trilogy of missions.

Microgravity could offer breakthroughs in drug development by enhancing solubility and bioavailability, which are key challenges in HIV and hepatitis C treatments.

VARDA
SPACE INDUSTRIES

"We had to design a maneuver sequence to optimize the flight path angle and velocity, and be as accurate as possible."

— Richard Hunter

Above: Earth reentry of the W-2 mission.

Opposite: The W-2 capsule reentry to the Koonibba Test Range in South Australia.

"ESCAPADE"

2025

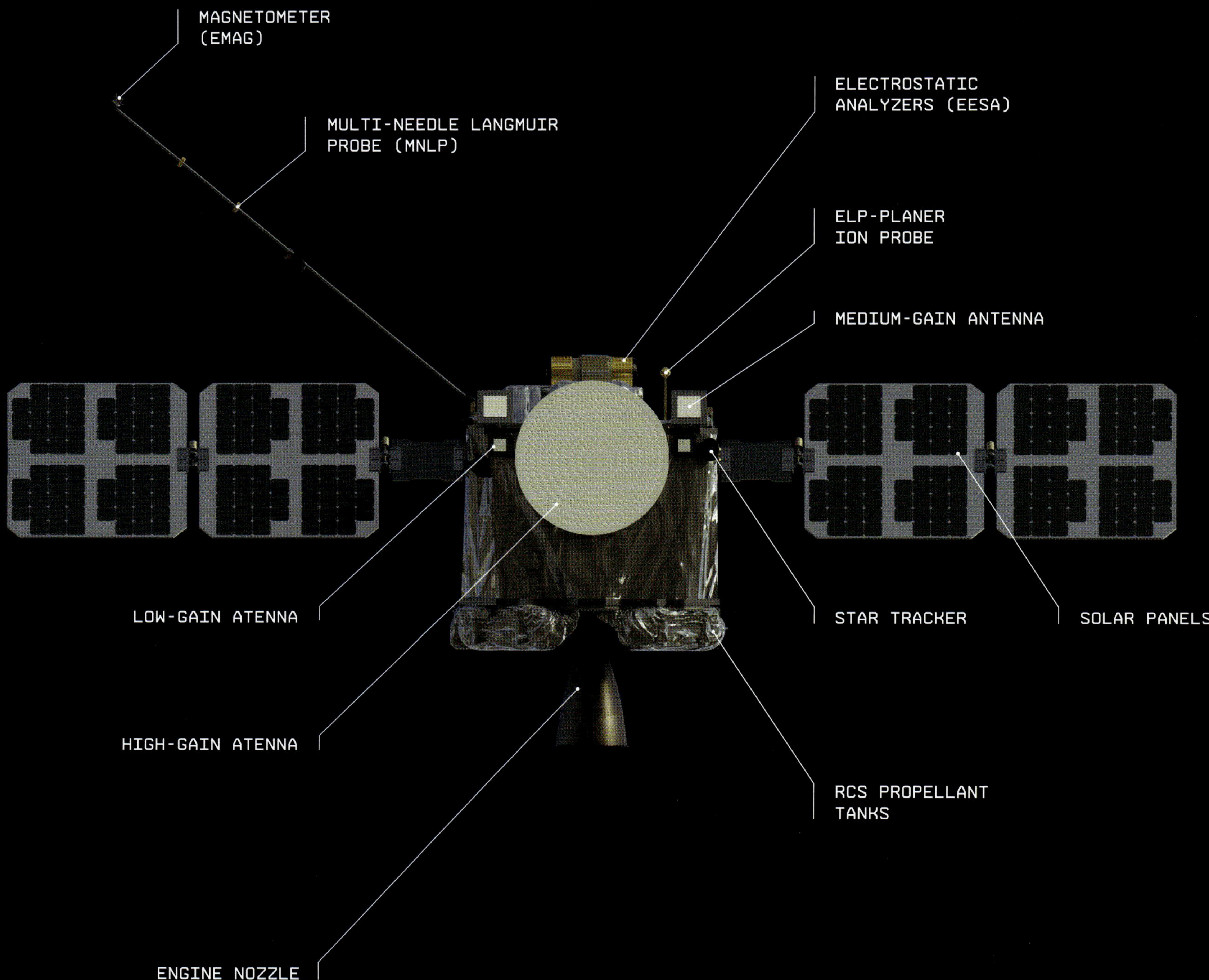

In 2024, Rocket Lab completed two spacecraft for delivery to Cape Canaveral, twin satellites that will undertake the “ESCAPADE” heliophysics mission for NASA and the University of California, Berkeley’s Space Sciences Laboratory.

The orbiters will study the atmosphere around Mars, gathering data that will inform efforts to protect astronauts and spacecraft from space weather as humans push further out into space, including to the Red Planet itself.

Once launched, the “ESCAPADE” mission will measure plasma and magnetic fields around the Red Planet. These observations will help scientists unravel the processes that strip atoms from Mars’s magnetosphere and upper atmosphere, offering critical insights into Martian climate evolution.

Known as Blue and Gold, the satellites were designed and assembled at Rocket Lab’s Spacecraft Production Complex and headquarters in Long Beach, California. “ESCAPADE’s” launch is constrained by the alignment of Earth and Mars, so a small delay can see the launch window missed. “ESCAPADE” is likely to launch in 2025 on Blue Origin’s New Glenn rocket.

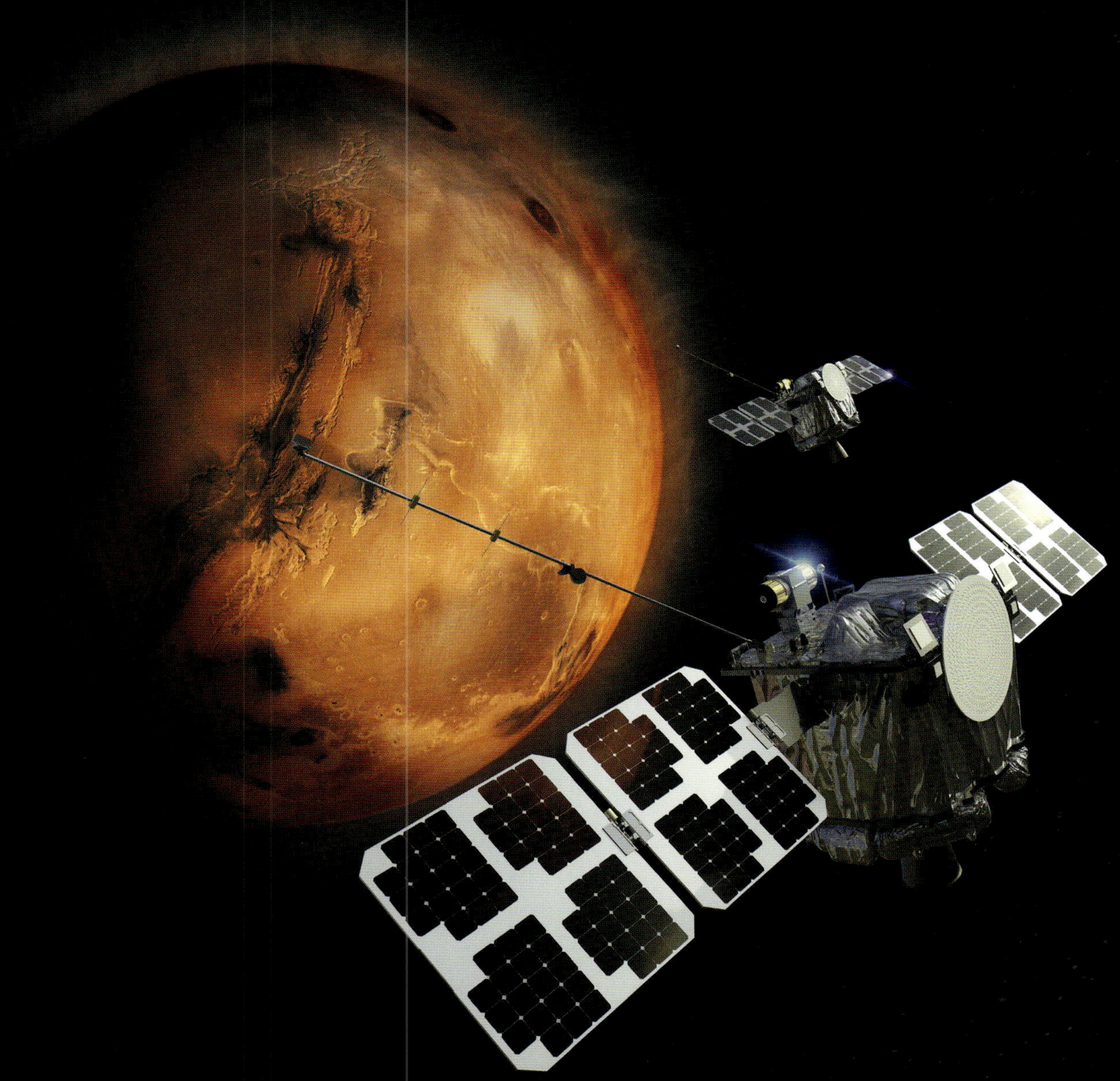

Previous page: A render of the two "ESCAPADE" spacecraft approaching Mars.

These pages and overleaf: Environmental testing of one of the twin spacecraft for NASA's "ESCAPADE" mission to Mars.

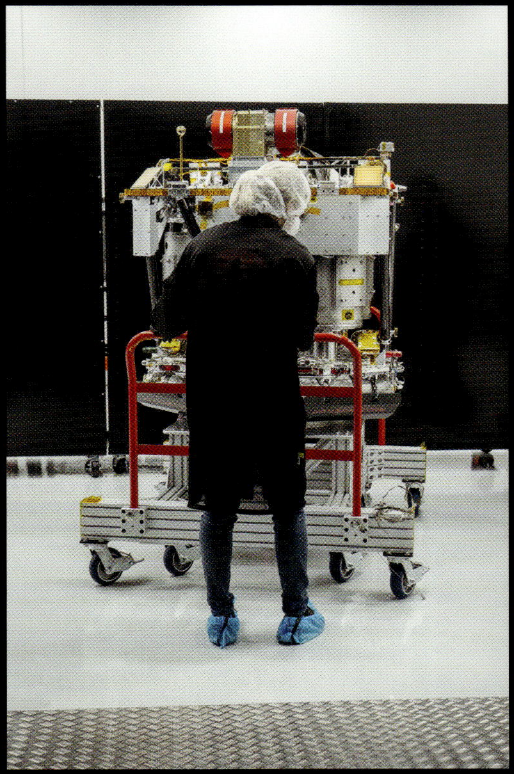

ROCKET

VENUS LIFE FINDER

2026–2027

"If you find life in the clouds of Venus, then you can pretty much assume that life is quite prevalent throughout the universe."

– Sir Peter Beck

NASA and other space agencies have undertaken over thirty missions to Venus, but Rocket Lab's upcoming mission will be the first private exploration of the planet.

"I want to see if there's life in the universe," said Peter Beck, who is so passionate about the Venus mission that he'd have Rocket Lab undertake it entirely on its own if there weren't any interested partners putting up their hands to work with the company on it.

Along with Massachusetts Institute of Technology, which is planning two missions under the Morning Star banner, Rocket Lab is sending the first private mission to Venus in search of supporting evidence of organic compounds in the cloud layer—traces of life.

The goal, using an Electron launch vehicle and Pioneer spacecraft, is to send a probe to around 30 miles' (48 km) altitude, where Venus's atmospheric conditions are closer to those found on Earth. The tiny probe, weighing less than 2.2 pounds (1 kg), will be dropped into the Venus atmosphere with a mission lifetime of around five minutes.

Venus has fascinated scientists for decades, largely because it was likely once a habitable planet similar to Earth. A 2019 study from NASA's Goddard Institute for Space Studies found that Venus could have had shallow oceans on the surface for two to three billion years, and this would have supported temperatures of between 68 to 122 degrees Fahrenheit (20 to 50°C).

Around 700 million years ago, though, a catastrophic geological event released carbon dioxide into the atmosphere, turning Venus into a dangerous, inhospitable planet where atmospheric temperatures reach 1,000 degrees Fahrenheit (538°C).

Phosphine (PH_3) was first detected in Venus's atmosphere in 2020 by a team led by Jane Greaves of Cardiff University, using both the James Clerk Maxwell Telescope in Hawai'i and the Atacama Large Millimeter Array in Chile.

The scientific team used the telescope to examine Venus's atmosphere for patterns of light that might indicate unexpected molecules and possible signatures of life. When observing Venus, they detected a specific absorption pattern in the light spectrum at a wavelength of about half an inch (13 mm), which matched what phosphine gas would emit within Venus's clouds.

This discovery sparked significant scientific interest, because phosphine is considered a "biosignature"—a molecule strongly associated with biological processes on Earth, where it's produced by microbes in oxygen-free environments.

"If you find life in the clouds of Venus, then you can pretty much assume that life is quite prevalent throughout the universe," Peter said.

Venus isn't the only place in the solar system to look for life, but Peter suggests it is the most promising. "I mean, Mars is dead and sterile; there's nothing there. Jupiter's Moon, Europa, has potential, but it's a very long way away."

This page: A synthesized false color image of Venus using 283-nm and 365-nm images taken by UVI photography.

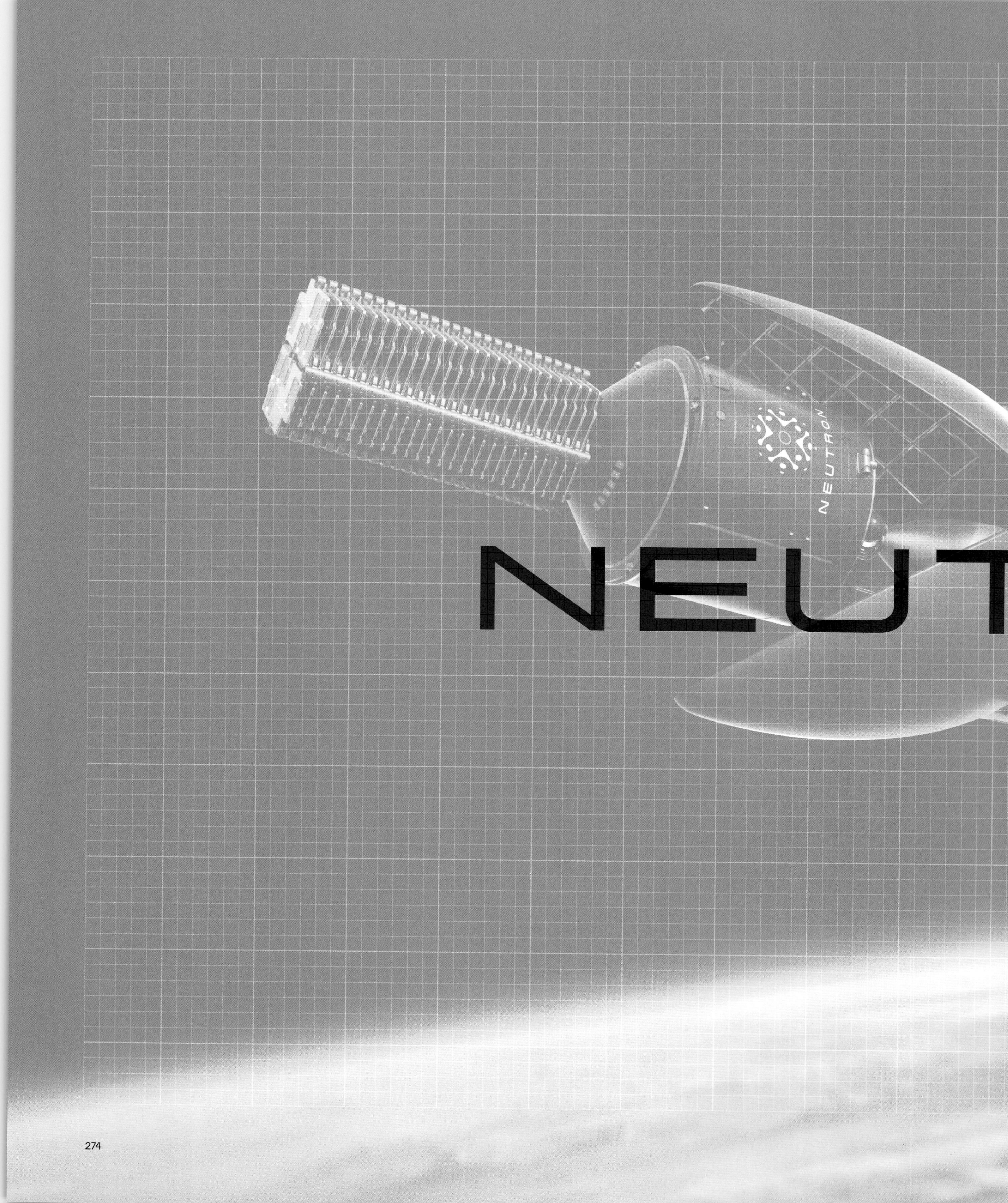

NEUT

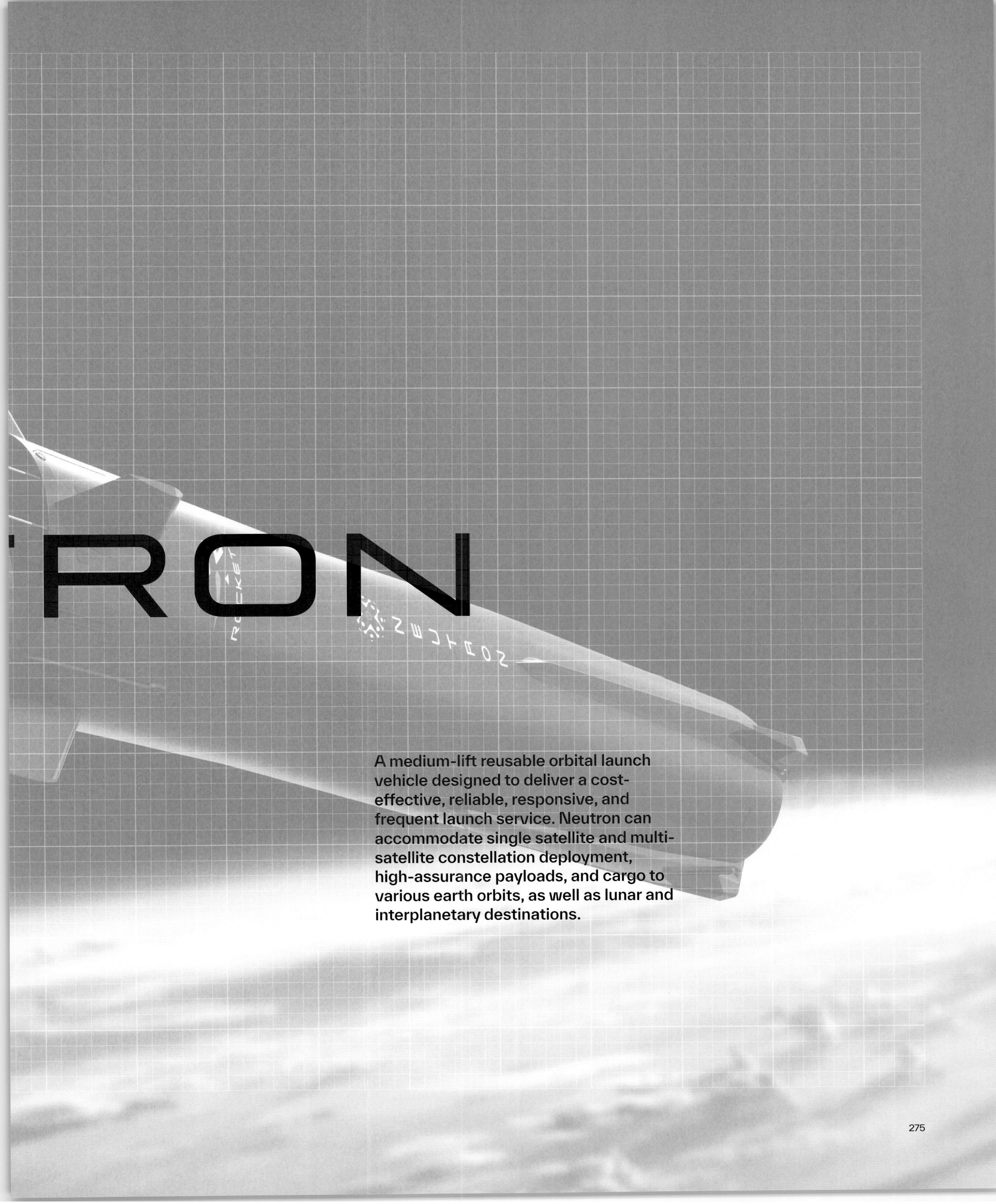

TRON

A medium-lift reusable orbital launch vehicle designed to deliver a cost-effective, reliable, responsive, and frequent launch service. Neutron can accommodate single satellite and multi-satellite constellation deployment, high-assurance payloads, and cargo to various earth orbits, as well as lunar and interplanetary destinations.

NEUTRON

NEUTRON

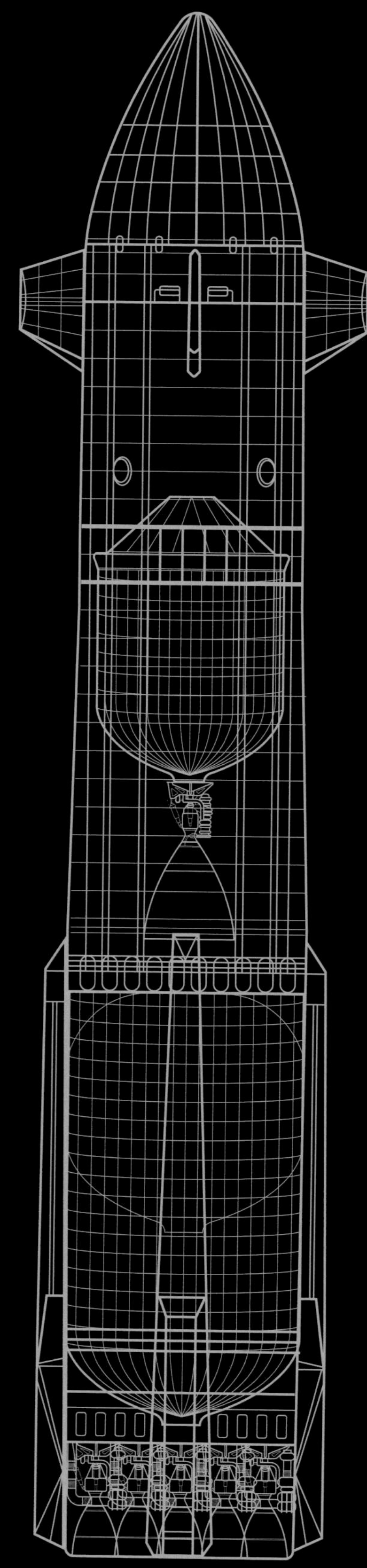

ROCKETLAB
NEUTRON

NEUTRON

FIRST STAGE

Neutron is a two-stage launch vehicle that stands 141 feet (43 m) tall with a 23-foot (7 m) diameter. Neutron features a fully reusable first stage designed to return to the launch site or land on an ocean platform, including a reusable fairing that remains attached to the launch vehicle's first stage when it returns to Earth. Its fairing diameter is 18 feet (5.5 m).

Neutron is powered by nine Archimedes engines on the first stage and a single vacuum-optimized relightable Archimedes engine on the second stage. The launch vehicle utilizes advanced carbon composite technology throughout many of its structures to provide additional performance, including its first and second stages and propellant tanks, making it the world's largest reusable carbon composite launch vehicle.

GENERAL SPECIFICATIONS	FIRST STAGE
Height	141 feet (43 m)
Diameter	23 feet (7 m)
Type	Pressurized structures—Monocoque
Material	Carbon Fiber Reinforced Polymer
PROPULSION	
Engine designation	Archimedes
Engine type	Oxidizer Rich Staged Combustion
Number of engines	9
Propellant	Liquid Oxygen (LOX) and Liquid Methane
Thrust (stage total)	6,600 kN, Sea Level
Propellant feed system	Turbopump
Throttle capability	65% Minimum
Restart capability	Yes
ASCENT ATTITUDE CONTROL	
Pitch, yaw	Gimbaled Engines
Roll	Gimbaled Engines
Coast attitude control	Cold Gas Thrusters

DESIGN INNOVATIONS

- Fully reusable first stage designed to return to launch site or land on an ocean platform
- Unique captive fairing design that remains attached to the first stage during return
- Estimated cost per launch: $50 million

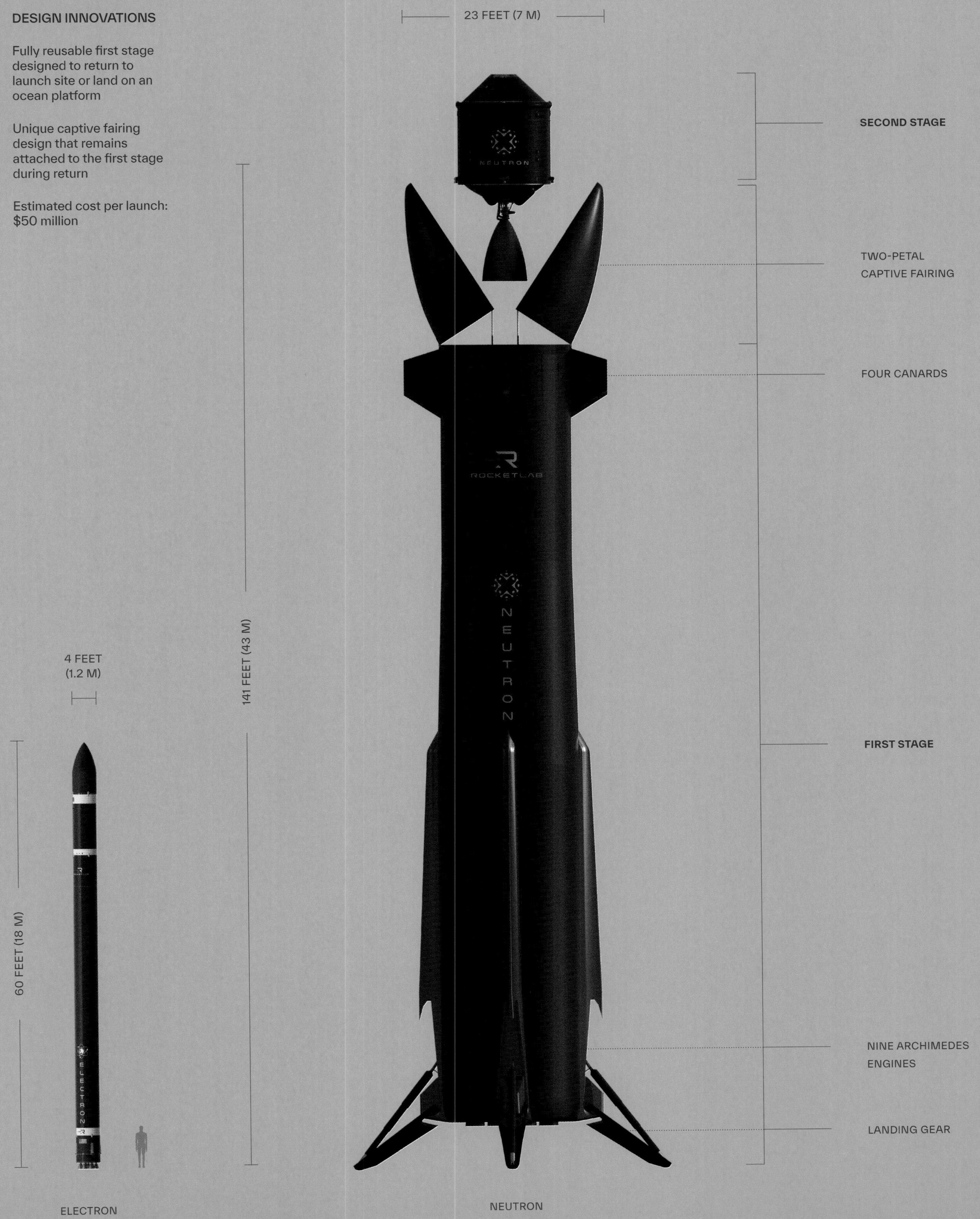

ELECTRON

NEUTRON

SECOND STAGE

Neutron's second stage (Stage 2) is made of carbon composite and consists of a single vacuum-optimized Archimedes engine and common bulkhead tanks for its LOX and methane propellants.

It is produced to be extremely mass-efficient and cost-optimized, as it is expended every launch, and is capable of multiple relights for constellation deployment and extended on-orbit operations.

Encapsulated and "hung" within Neutron's Stage 1, the Stage 2 is suspended by its top flange inside the interstage at a rigid mounting interface. In addition to providing easily accessible and condensed mounting location for avionics hardware, aerodynamic control devices, and fluids lines, this interface also minimizes the requirement for the Stage 2 to withstand the external launch environment during ascent.

Driven by a pneumatic impulse delivery system, Neutron's Stage 2 separates quickly to maintain high performance while keeping shock and acceleration loads to a minimum.

GENERAL SPECIFICATIONS	SECOND STAGE
Height	37.75 feet (11.5 m)
Diameter	16 feet (4.9 m)
Type	Pressurized structures—Monocoque
Material	Carbon Fiber Reinforced Polymer

PROPULSION	
Engine designation	Archimedes Vacuum
Engine type	Oxidizer Rich Staged Combustion, Vacuum-Optimized
Number of engines	1
Propellant	Liquid Oxygen (LOX) and Liquid Methane
Thrust (stage total)	900 kN, Vacuum
Propellant feed system	Turbopump
Throttle capability	65% Minimum
Restart capability	Yes
ASCENT ATTITUDE CONTROL	
Pitch, yaw	Gimbaled Engine
Roll	Cold Gas Thrusters
Coast attitude control	Cold Gas Thrusters

PAYLOAD CAPACITY

→ 14.3 tons (13,000 kg) to Low Earth Orbit (LEO) with reusability

→ 16.5 tons (15,000 kg) in expendable configuration

→ 9.4 tons (8,500 kg) when returning to launch site

→ Up to 2.2 tons (2,000 kg) to the Moon on trans-lunar injection

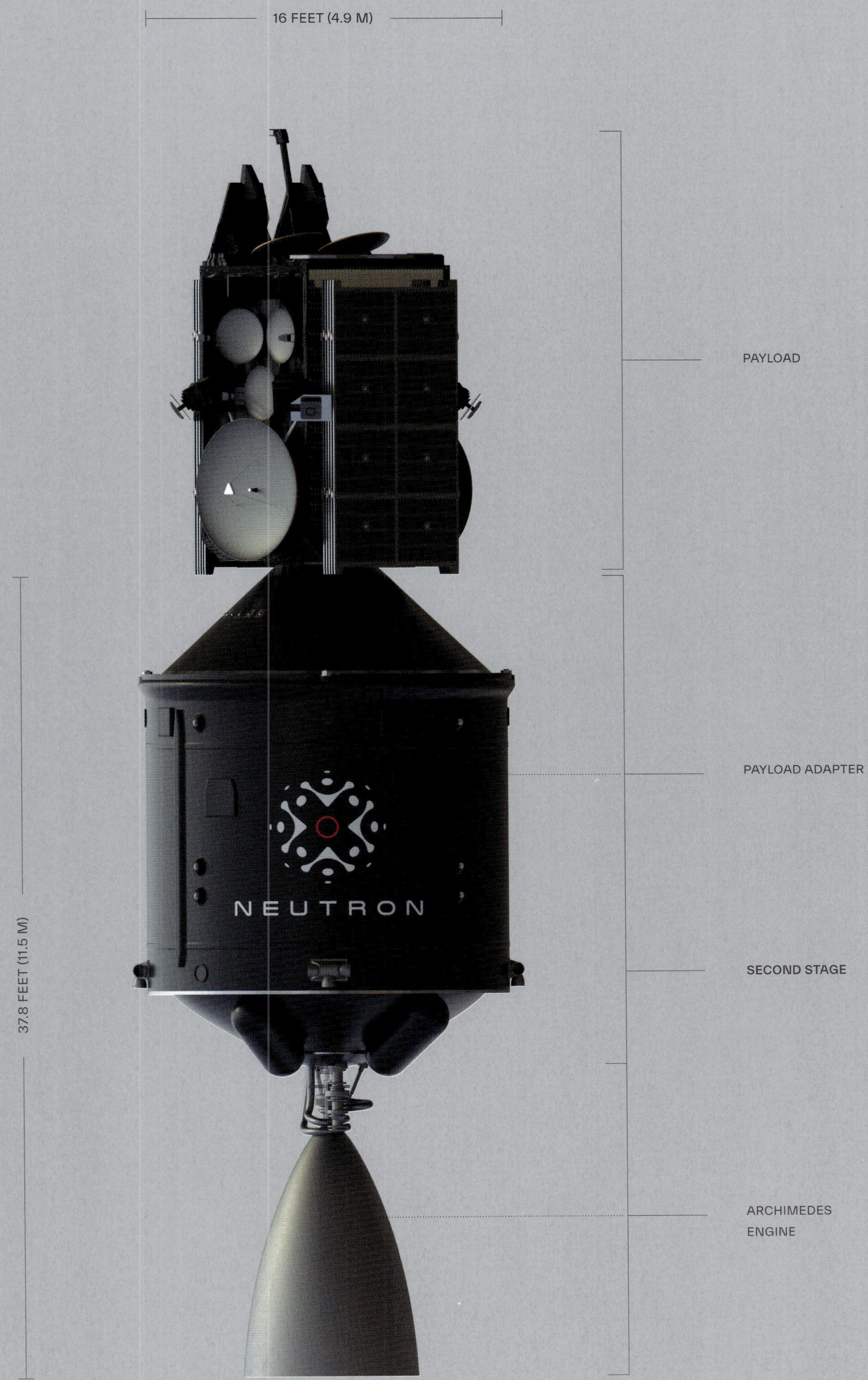

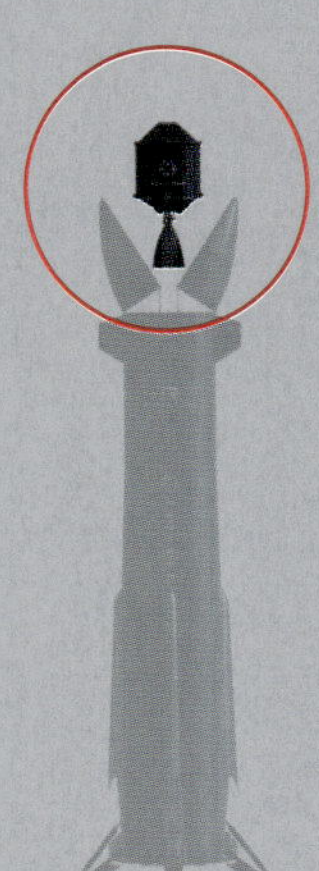

“The company is ready to step up to answer the call to explore the solar system.”

— Sir Peter Beck

NEUTRON

Neutron: A Constellation of Possibilities

It was a warm and still afternoon in Hancock County, Mississippi, when Archimedes first roared into life.

The first static hot-fire test of the engine that will power the medium-lift Neutron rocket went off without a hitch at NASA's Stennis Space Center in 2024. A fiery tail extended nearly 100 feet (30 m) behind Archimedes, a rainbow of colors shimmering in the billowing exhaust.

It was a first-stage version of Archimedes, nine of which will be attached to the Neutron rocket to deliver a total of 1,450,000 pound-force (6,660 kN) of thrust.

"From here it is about dialing the engine in, building a bunch more of them, and getting them rolling off the production line," Peter Beck told *Space News* after the hot-fire.

On the same day, 495 miles (796 km) away at Rocket Lab's Space Structures Complex in Middle River, Maryland, the former site of Lockheed Martin's vertical launch building, a giant piece of equipment was being installed on the future Neutron production line.

The automated fiber placement (AFP) machine is what Rocket Lab uses to make Neutron's carbon composite rocket components, from its second-stage propellant tank, to the fairings and panels that make up its first stage and interstage. Neutron will be the largest carbon composite rocket ever made, and the automation involved in its manufacture makes producing Electron look like an artisanal endeavor in comparison.

The man overseeing development of Neutron, Shaun D'Mello, was integrally involved in the genesis of Electron. He was taken aback when Peter shoulder-tapped him in 2021 to head the company's largest and most ambitious project.

"There was definitely no other answer other than 'Yes,'" he figured. "Having gone through Electron and come out the other side, I was ready to do it at least one more time, one more rocket program from start to finish."

Shaun's first task was to assemble a propulsion team to work on Archimedes. The engine parts would need to be 3D-printed in a way that ensured Archimedes' reusability, with a minimum reuse target of up to twenty launches per engine. The nine engines needed to produce enough thrust combined to lift 28,660 pounds (13,000 kg) of payload to space.

"The criteria to join the Archimedes team was that you had to have worked on an engine this big before," said Shaun. "We were able to pull some really great people who had worked at SpaceX, Blue Origin and others, where they had tackled this problem before," he added.

The difference was that other medium-lift rocket engines typically had a billion-dollar development budget, and took up to a decade to get into production. The entire budget to bring Neutron to market was around $300 million.

Shaun was elated when the Archimedes hot-fire went so well, reaching 102 percent power without skipping a beat. More often than not in this industry, you can expect an engine to explode on the test stand on its first hot-fire, or even fail to ignite.

But with an ambitious deadline for Neutron's maiden launch in late 2025, the race was on to marry the Archimedes engines with the Neutron rocket body and perfect the avionics and control systems and software—everything the Rocket Lab team had done with Electron, but on a larger scale.

Perfecting the Hungry Hippo fairing design is also critical to Neutron's success. Unlike traditional rockets that jettison fairings, Neutron's two clamshell composite fairing panels open in space to deploy the second stage and payload, then close back up and remain attached to the first stage for reentry.

This design eliminates the need for complex fairing recovery systems. What lands back on the pad looks externally exactly the same as what lifted off—a complete rocket—not the norm in the industry.

Below, pages 288–89 and 292–93: Test firing Rocket Lab's Archimedes engine.

Pages 274–79, and 285: Renders of Rocket Lab's Neutron rocket.

ARCHIMEDES

Neutron is powered by the Archimedes engine, an oxidizer rich staged combustion cycle engine that will power the reusable Stage 1 and expendable Stage 2 of the launch vehicle.

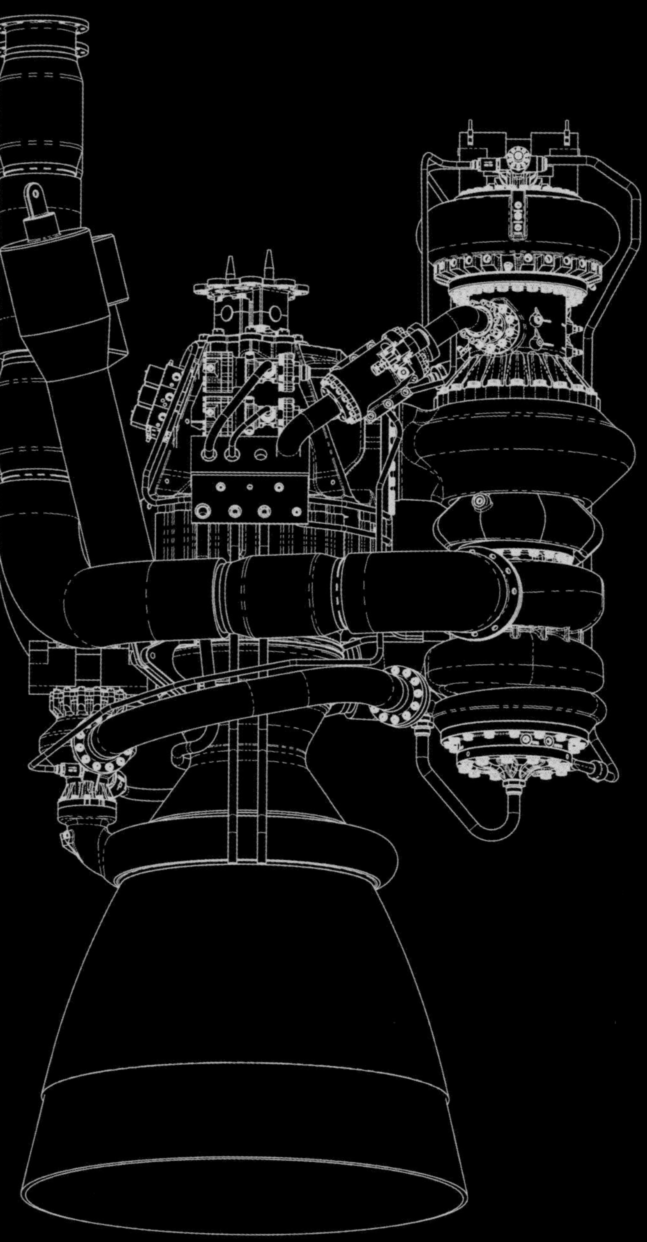

Capable of producing up to 165,000 pounds (734 kN) of thrust (lbf) per engine, Archimedes operates at lower stress levels than other rocket engines to enable rapid and reliable reusability. The total combined thrust (lbf) of nine Archimedes engines for Neutron's Stage 1 is designed to reach 1,450,000 pounds (6,450 kN). Each engine burns liquid methane and liquid oxygen, which provides higher specific impulse than RP-1 kerosene for improved stage performance without the dramatic stage size increase required for low-density liquid hydrogen.

→ 165,000 pounds (734 kN) of thrust (lbf) per engine.

→ Nine Archimedes engines produce a total combined thrust (lbf) of 1,450,000 pounds (6,450 kN).

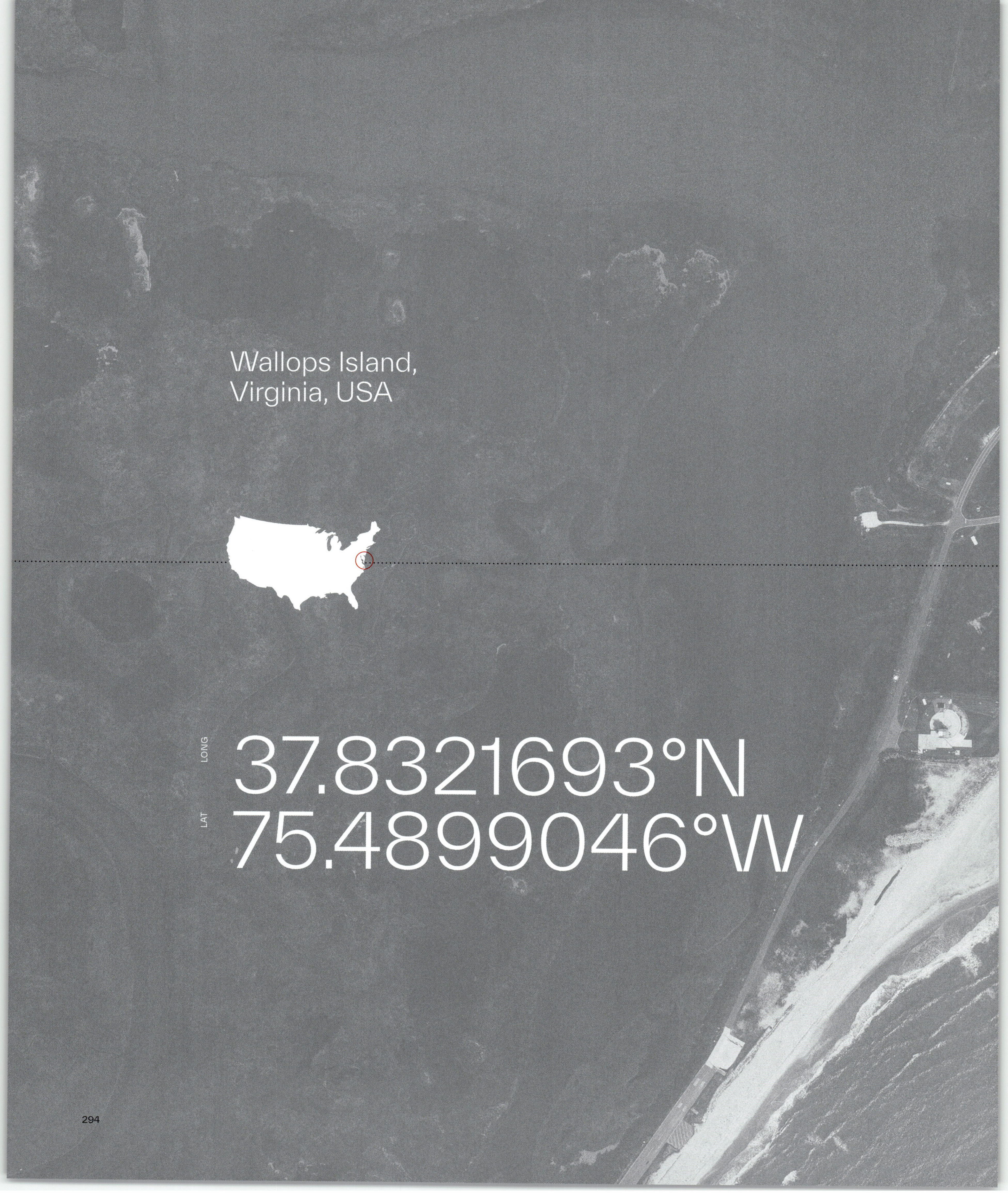

Wallops Island,
Virginia, USA

LAT 37.8321693°N
LONG 75.4899046°W

LAUNCH COMPLEX: 3

Launch Complex 3 is the dedicated launch site for Rocket Lab's new reusable rocket Neutron. Located at the Mid-Atlantic Regional Spaceport within NASA's Wallops Flight Facility, Neutron's launch site is located right next door to Launch Complex 2. As a reusable rocket, Neutron's first stage can support return-to-launch-site and downrange landing options, before return to Launch Complex 3 for refurbishment and reflight.

END T

O END

This page: A rendering of Rocket Lab's Neutron rocket.

Breaking the Medium-Lift Monopoly

Shaun D'Mello: Vice President—Neutron, Rocket Lab said the ultimate challenge of developing a next-generation launch vehicle that could meet Rocket Lab's grand ambitions for space was what motivated him.

> "Clearly there's a launch monopoly in medium-class launch. No monopoly ever survives the test of time. We are probably the most likely to be able to have a crack at actually breaking it up."
>
> — Sir Peter Beck

"The biggest thing that gets me up in the morning is the fact that we're building a rocket that can lift and put Rocket Lab's own infrastructure up there too," he explained. "We're building our own satellites, launching them on Neutron, then being able to beam the services back down. We couldn't do that with Electron."

Peter put it in even simpler terms: "Clearly there's a launch monopoly in medium-class launch," he said of SpaceX, the only company currently launching regular medium-lift rockets carrying satellites to orbit. "No monopoly ever survives the test of time. We are probably the most likely to be able to have a crack at actually breaking it up."

Not only would Neutron allow Rocket Lab to deliver larger and more numerous satellites to orbit, but it would enable Rocket Lab itself to form its own satellite constellation and deploy its own satellites frequently on Neutron.

"I firmly believe that in five to ten years' time, all of the big space companies on this planet will be end-to-end space companies looking exactly like Rocket Lab," said Peter.

While launching larger satellites and building constellations of them is the primary intended use of Neutron, its larger capacity opens up other possibilities too.

"The International Space Station is going to be retiring soon," Shaun pointed out. "There's an interesting mission in building and deploying commercial infrastructure in space such as space stations and refueling depots, a side of the industry that's just never existed before," he said.

Then there are the interplanetary missions on Rocket Lab's radar, including its potential involvement in the Mars Sample Return mission, if its design concepts impress NASA, and the mission to Venus.

"Now it becomes all that more possible, because we've got a rocket that's just that much more capable to go and do it," Shaun concluded.

The economics of the space industry dictated that Rocket Lab had to move beyond Electron to offer larger launch capacity, while developing the capability in Space Systems.

"If you look at the space industry, launches are worth about ten billion a year, satellites are twenty to thirty billion a year," said Peter. "The applications, or infrastructure in orbit, is worth 320 billion."

For Peter, Neutron is the natural next step in Rocket Lab's journey. Remaining as a highly successful small rocket company just wasn't an option.

"It all goes back to what I was told as a kid, which is to have the biggest impact you can have," he said. "It might be infrastructure in orbit, or pulling some samples back from Mars and attempting to answer the biggest questions about the universe. That's an enduring impact. There are certain things where you just have to go and do it."

No Manhattan Project Required

Likewise, Rocket Lab's interest in NASA's project to scope out options for returning samples gathered on the surface of Mars demonstrates Peter's appetite for complex projects aimed at expanding our understanding of the universe.

"It's arguably the most complex interplanetary mission since they landed on the Moon," he said. "But there's no gap in the engineering. There's no Manhattan Project required to execute it. It's literally stuff that we've all done before. You've just got to do them all at once."

Which is really the story of Rocket Lab as it approaches its twentieth anniversary. It's a company diversifying its offerings, investing in advanced technologies, and pursuing strategic acquisitions.

Rocket Lab is laying the groundwork for a future where it looks likely to play a pivotal role in the shape of the commercial space sector. Despite the years of hard work that have gone into the company, it's a mission that's really only just begun.

“Piecing The Whole Thing Together”— The Future of Rocket Lab

“I say this every year, but it feels like this year is a pretty important one for Rocket Lab,” Sir Peter Beck remarked in early 2025 after an earnings call that updated the market on the status of Neutron, and outlined the company’s ambitions to develop its own satellite constellations.

“One of the common misconceptions we’re faced with is that Rocket Lab is just a launch company. Yes, launch is something we’re an industry leader in, and of course Rocket being in the name doesn’t help, but it’s just a part of what we do,” said Peter.

Peter is ambitious about transforming Rocket Lab into an end-to-end company which, in the space industry, means becoming the one-stop shop that supplies everything needed for a mission to space: from the smallest component that makes up a satellite and the software that runs it, to the rockets that launch it, and the management and control of that mission once it’s in space. Being across all three is a potentially lucrative business.

An End-to-End Space Company
What every launch, every development milestone, and every spacecraft is building toward for Rocket Lab’s future.

The Ride to Space
To do important things in space, first you have to get there, and Electron and Neutron are the rockets to do that. With small satellites covered by Electron and entire constellations and large spacecraft ideal for launch on Neutron, Rocket Lab has much of the space industry covered when it comes to satellite deployment.

The Tools in Space
Next, the tools to do things in space once you get there. This is the spacecraft, the critical components that make them function, the software, the ground systems, and all the other pieces that make missions possible in orbit. All of which are functions and machines Rocket Lab has already developed and built, delivering spacecraft and missions for commercial and government satellite operators globally.

"It all goes back to what I was told as a kid, which is to have the biggest impact you can have."

— Sir Peter Beck

The whole reason of going to space in the first place is to conduct science, gather data, and deliver services from space that impact people around the world—with much of that provided by the satellites and spacecraft that are carried by rockets to space.

By owning rockets and spacecraft themselves, and with the expertise to command both at the same time, Rocket Lab is gearing up to provide that science, that data, and those services themselves with their own constellation in the future. There's no waiting in line for a rocket when you control your own launch schedule, and you're not dependent on someone else or the supply of components and materials when you can design, build, test, and operate your own spacecraft.

What Rocket Lab's future constellation will be, only time will tell. The company isn't revealing what it has in mind just yet. But as Rocket Lab knocks on the door of the debut launch of Neutron (a rocket that can carry dozens of satellites to space at once, and launch over and over again as a reusable launch vehicle), reveals new satellite architectures like Flatellite, and takes on more missions that expands the capabilitites of what the company can do—clues can be found if you know where to look.

Outsized ambition is really the story of Rocket Lab as it approaches its twentieth anniversary. It's a company diversifying its offerings, investing in advanced technologies, and pursuing strategic acquisitions.

Rocket Lab is laying the groundwork for a future where it looks likely to play a pivotal role in the shape of the commercial space sector. Despite the years of hard work that have gone into the company, it's a mission that's really only just begun.

"Everyone thinks you are crazy until you do it; then you are just called a visionary."

— Sir Peter Beck

Glossary of Terms

3D printing	Also known as additive manufacturing. A manufacturing process where objects are built layer by layer from digital models, often using metal or plastic. In rocketry, it allows for complex engine parts to be produced more efficiently than traditional methods.
Aerospace	The atmosphere and space beyond; or the sector involved in aviation and space flight.
AFP	Automated fiber placement: A manufacturing technique where carbon fiber is laid down by a machine in precise patterns to create strong, lightweight composite structures.
AFTS	Autonomous flight termination system: A safety system that can automatically terminate a rocket's flight if it veers off course, without human intervention.
AIM-9 Sidewinder	A type of air-to-air missile used by military aircraft, referenced in the context of propulsion technology development.
Archimedes	A 3D-printed, reusable rocket engine used to power Rocket Lab's Neutron rocket.
Artemis	Refers to NASA's Artemis Moon exploration program, which aims to establish a permanent base on the Moon to enable human missions to Mars.
Ātea	Rocket Lab's first suborbital rocket, Ātea-1. Ātea is the te reo Māori word for space.
Attitude control	Systems onboard spacecraft that autonomously control its orientation relative to a target reference frame.
Avionics	Electronics used on a spacecraft.
"CAPSTONE"	A microsatellite testing the calculated orbital stability of a Near Rectilinear Halo Orbit (NRHO) around the Moon. It was launched aboard an Electron rocket in 2022.
Carbon composites	Carbon fiber-reinforced polymers (CFRP): Advanced materials made from carbon fibers embedded in a polymer matrix, offering high strength and low weight—ideal for rocket structures.
CubeSat	Small satellites used for space exploration and research.
Curie	A 3D-printed engine used to power various Rocket Lab spacecraft and Electron's Kick Stage.
DARPA	The Defense Advanced Research Projects Agency, part of the United States Department of Defense.
Delta-V	The change in velocity, or amount of velocity, needed for spacecraft to perform certain maneuvers.
Downrange	The horizontal distance a spacecraft travels from its launch site.
ELaNa	Educational Launch of Nanosatellites: A NASA program that launches small satellites built by educational institutions.
Electron	A partially reusable orbital launch vehicle first launched in 2017.
"ESCAPADE"	NASA's Escape and Plasma Acceleration and Dynamics Explorers mission, which aims to measure magnetic fields around the Red Planet. It is the first multi-spacecraft orbital science mission to Mars.
FAA	Federal Aviation Administration: The US government agency that regulates civil aviation, including licensing rocket launches.
Fairing	The protective nose cone of a rocket that shields the payload during launch and is jettisoned once in space.
Flatellite	A flat satellite designed for mass production and customization for extensive satellite constellations.
Flywheel	A mechanical device that stores rotational energy.
G-force	Also known as "g's." A measure of acceleration felt as weight; 1 g is equivalent to the force of Earth's gravity.
Heliophysics	The study of the Sun and its influence on the solar system.
HyperCurie	A high-performance spacecraft engine developed by Rocket Lab for the "CAPSTONE" and "ESCAPADE" missions to the Moon and Mars.
Hypersonic	Speeds greater than five times the speed of sound.
Interplanetary	Travel between planets.
ITAR	International Traffic in Arms Regulations: US government regulations controlling the export of defense-related technology, including rocket designs.
Kármán Line	The internationally recognized boundary of space, located 62 miles (100 km) above sea level.
Kick Stage	A stage used to provide extra velocity to a payload after it has been deployed from the Electron launch vehicle.
Launch complex	A purpose-built facility for the design, manufacturing, and launch of spacecraft.
Launch vehicle	A rocket-powered vehicle designed to transport payloads beyond Earth's atmosphere.
LC-1, LC-2, LC-3	Launch Complex 1, 2, and 3. Rocket Lab's dedicated launch sites: LC-1 in Mahia, New Zealand; LC-2 and LC-3 at Wallops Island, Virginia, US.
Lift-off thrust	The force that enables a rocket to overcome gravity and begin its ascent into space.
Low Earth Orbit	An orbit around Earth at an altitude of 1,200 miles (1,931 km) or less, often referred to as LEO.
LOX	Liquid oxygen. A cryogenic liquid used as an oxidizer in rocket engines.
Mach	A unit of speed relative to the speed of sound; Mach 1 is the speed of sound; Mach 5 is five times the speed of sound.
Microgravity	Weak gravity which causes people or objects to appear weightless.

Mission Control	An Earth-based facility that manages spaceflights.
MTCR	Missile Technology Control Regime: An international agreement aimed at limiting the spread of missile and rocket technology capable of delivering weapons of mass destruction.
NASA	The National Aeronautics and Space Administration, US.
Neutron	A reusable medium-lift launch vehicle for the delivery of payloads to multiple orbits. It is expected to be the largest carbon composite rocket ever made.
Nitrous oxide	A chemical compound used as an oxidizer in some rocket propellants.
Orbital inclination	The angle between a satellite's orbital plane and the equator, determining the path it takes around Earth.
Orbital lifespan	The duration of time an object stays in orbit before reentering Earth's atmosphere or being moved out of orbit.
ORS	Operationally Responsive Space Office: A US Department of Defense unit focused on rapid deployment of small satellites.
Overview effect	A state of awe at the Earth's fragility experienced by astronauts viewing Earth from space.
Payload	The object or objects carried by a launch vehicle into space to complete a mission, e.g. a satellite.
Photon	The original Rocket Lab spacecraft, Photon is an integrated launch-plus-spacecraft solution with a modified Electron Kick Stage.
Propellant	A chemical mixture of a fuel and oxidizers that is burned to produce thrust and power rockets.
PID Controller	Proportional-Integral-Derivative Controller: A type of control algorithm used in engineering to maintain desired system performance, such as rocket trajectory.
Radio frequency	A measurement representing the oscillation rate of electromagnetic radiation spectrum.
Reaction Control System	Small thrusters that provide precise attitude control.
Reaction wheels	An electric motor attached to a flywheel that allow satellites to be repositioned while in orbit.
Red Planet	Another name for Mars.
Rocket fuel	Also referred to as propellant. The substance used to power rockets.
Rover	A robotic vehicle controlled from Earth and designed to move across the surface of a planet.
Rutherford	A small liquid-propellant rocket engine developed by Rocket Lab that powers its Electron rocket. It is the first 3D-printed rocket engine to reach space, and the first rocket engine to use battery-powered electric fuel pumps.
Satellite	A man-made device placed in orbit around Earth or the Moon to collect and communicate data.
Separation systems	Mechanisms used to ensure payloads remain connected to launch vehicles during launch, then are safely deployed in space.
Solar arrays	Large-scale solar panels deployed in orbit around the Earth to gather solar energy and convert it into electricity.
Solar cells	An electronic device that converts sunlight into electricity to power spacecraft and satellites, often referred to as a photovoltaic cell (PV cell).
Solar phenomena	Natural phenomena occurring within the Sun's atmosphere and taking various forms including solar wind, radio wave flux, and solar flares.
SDA	Space Development Agency: A US Space Force unit tasked with deploying disruptive space technology.
SpaceX	Space Exploration Technologies Corporation, a US aerospace company established by Elon Musk in 2002.
Spacecraft	A vehicle designed for travel and/or operation outside of Earth's atmosphere.
Spacecraft bus	Also known as a satellite bus, this is the support system of a spacecraft, providing essential functions for the payload and enabling missions to go ahead.
Specific impulse	A measure to determine the thrust of a rocket.
Stage separation	The process where one section of a multi-stage rocket detaches after its fuel is spent, allowing the next stage to ignite.
Stack testing	The operation of firing up an Electron rocket's engines as though it is about to launch, while keeping the rocket held down to the test mount.
Star tracker	Optical sensors that help satellites maintain their orientation and location in space.
Starlink	A satellite internet constellation developed by SpaceX.
Suborbital	A spaceflight that reaches outer space but does not have the speed to complete a full orbit around Earth and instead falls back to the surface.
Sun-Synchronous Orbit	SSO: A type of polar orbit that allows a satellite to pass over part of Earth at local solar time.
T-minus	The countdown indicating how much time remains before a scheduled space launch.
TSA	Technology Safeguards Agreement: A bilateral agreement allowing the transfer and secure management of sensitive rocket technology between countries.
Turbopump	A high-speed pump that delivers propellant to the combustion chamber, enabling the engine to produce thrust.
United States Space Operations Command	Part of the US Department of Defense responsible for the operational employment of space forces.
Varda Space Industries	A US research company that processes materials in orbit and returns them to Earth.
Viscous Liquid Monopropellant	A single-part high-density propellant that simplifies a rocket's propulsion system, also known as VLM.

Index

Note: Page numbers in italics denote a photograph or illustration

Acknowledgments

We are tremendously grateful to everyone who contributed to the making of this book.

Firstly, we would like to thank all those individuals who generously gave up their time to be interviewed: Yesica Baeza, Sir Peter Beck, Desirée Brundin, Dr. Brad Clevenger, Morgan Connaughton, Christian Davenport, Shaun D'Mello, Alek Derejian, James Dreyer, Richard French, Katie Gwozdecky, Richard Hunter, Steven Joyce, Vinod Khosla, Jane McNicoll, Ben Malcolm, Lachlan Matchett, Shaun O'Donnell, Michael Pearson, Brian Rogers, Dr. Sandy Tirtey, and Julia Wembacher.

To the entire Rocket Lab team, but especially Sir Peter Beck, Morgan Connaughton, Murielle Baker, Adam Herlihy, and James Rattray, without whom this book would not have been possible, thank you. And last but not least, our gratitude to the editorial and design team: Peter Griffin who conducted the interviews and created the text; Cameron Gibb at Project Basis for book design; Nikki Addison for additional research, editing, and fact-checking; Mike Wagg for proofreading; Judith Wilson for indexing; and Blackwell & Ruth principals Geoff Blackwell and Ruth Hobday.

Additional image credits:
Front and back endpapers, pages 10–11, 20–21, 50–51, 184–85, 270–71 Mars surface; pages 2, 22–23 Electron rockets launching; pages 4, 223 Ring Nebula; pages 8–9 NGC 2683 galaxy; pages 5, 12, 14–15, 301, 306–7 Electron rockets launch at night; page 6 Mars; page 17 Sir Peter Beck; pages 18–19, 54–55, 58–59, 70, 164–65, 168, 169 Electron rocket engines firing on launch; 52–53 Electron rocket engine from below; pages 68–69, 71 Electron rockets in Rocket Lab's manufacturing facility; pages 142–43, 294–95 Wallops Island, Virginia, US; page 229 a view of the Sun; pages 234–35, 236–37 a view of the Moon; pages 304–5 NGC 2336 galaxy.

First published in 2025 by Blackwell and Ruth Limited, PO Box 37692, Newmarket, Auckland 1023, New Zealand, blackwellandruth.com

Design and layout by Cameron Gibb, Project Basis for Blackwell and Ruth Limited.

ISBN: 978-0-473-74122-8
10 9 8 7 6 5 4 3 2

A catalogue record for this book is available from the National Library of New Zealand. Kei te pātengi raraunga o Te Puna Mātauranga o Aotearoa te whakarārangi o tēnei pukapuka.

Distributed in the USA and Canada by ABRAMS, 195 Broadway, 9th Floor, New York, NY 10007, USA, abramsbooks.com

ABRAMS is represented in the UK and Europe by Abrams & Chronicle Books, 1 West Smithfield, London EC1A 9JU and Média Participations, 57 rue Gaston Tessier, 75166 Paris, France.
abramsandchronicle.co.uk and media-participations.com
info@abramsandchronicle.co.uk

Abrams books are available at special discounts when purchased in quantity for premiums and promotions as well as fundraising or educational use. Special editions can also be created to specification. For details, contact specialsales@abramsbooks.com or the address below.

Manufactured in China by 1010 Printing Ltd.

This book is made with FSC®-certified paper and other controlled material and is printed with soy vegetable inks. The Forest Stewardship Council® (FSC®) is a global, not-for-profit organization dedicated to the promotion of responsible forest management worldwide to meet the social, ecological, and economic rights and needs of the present generation without compromising those of future generations. The shrink-wrap is one hundred per cent recyclable via certified soft-plastics recycling programs.

FSC
www.fsc.org
MIX
Paper | Supporting responsible forestry
FSC® C016973